CYCLIC HOMOLOGY OF ALGEBRAS

CYCLIC HOMOLOGY OF ALGEBRAS

Peter Seibt

CNRS, Centre de Physique Theorique

Marseille, France

Published by

World Scientific Publishing Co. Pte. Ltd.
P.O. Box 128, Farrer Road, Singapore 9128

U. S. A. office: World Scientific Publishing Co., Inc.
687 Hartwell Street, Teaneck NJ 07666, USA

Library of Congress Cataloging-in-Publication data is available.

CYCLIC HOMOLOGY OF ALGEBRAS

ISBN 9971-50-468-5
9971-50-470-7 pbk

Printed in Singapore by Utopia Press.

To Sebastian, Julie

Anna and Willy

TABLE OF CONTENTS

Introduction. ix

Chapter I. **CYCLIC (CO)HOMOLOGY AND HOCHSCHILD (CO)HOMOLOGY.** 1

I.1 **Preliminaries: Spectral Sequences.** 1
I.1.1 Filtered Complexes and Exact Couples. 1
I.1.2 The Spectral Sequence associated with an Exact Couple. 3
I.1.3 Convergence of a Spectral Sequence. 7
I.1.4 Double Complexes and their Spectral Sequences. 12

I.2 **Cyclic (Co)homology and Hochschild (Co)homology.** 17
I.2.1 The Double Complex C(A). 17
I.2.2 The Cyclic Homology of an Associative Algebra. 20
I.2.3 Generalities about Mixed Complexes. 23
I.2.4 Cyclic Homology and Hochschild Homology. 32
I.2.5 Nonunital and Reduced Cyclic Homology. 41
I.2.6 Cyclic Cohomology. 58
I.2.7 Morita-Invariance of Hochschild Homology and of Cyclic Homology. 72

Comments on Chapter I. 84

References to Chapter I. 85

Chapter II. **PARTICULARITIES IN CHARACTERISTIC ZERO.** 86

II.1 **Relation to de Rham Theory.** 86
II.1.1 A First Approach: Noncommutative de Rham Complexes. 86
II.1.2 Cyclic Homology and de Rham Cohomology of Commutative Algebras. 104

II.2	**Relation to Lie Theory.**	121
II.2.1	Preliminaries around Invariant Theory.	121
II.2.2	Cyclic Homology and the Lie Algebra Homology of Matrices.	133
	Comments on Chapter II.	156
	References to Chapter II.	157
Further references.		158
List of symbols and notations.		159
Index.		160

Introduction.

These lectures are an extended version of my contribution to a seminar on cyclic cohomology, held at the University of Marseille Luminy, in 1985. They are essentially based on a paper of J.L. Loday and D. Quillen: Cyclic homology and the Lie algebra homology of matrices (Comment. Math. Helvetici, 59 (1984), 565-591), and contain also ideas and results of M. Karoubi and C. Kassel.

The exposition is purely algebraic, according to my own background, and thus concentrates rather on cyclic homology (than on cohomology), the former being a more natural starting point for an algebraist. But many of the leading ideas of the theory, more apparent in cohomology, come from topology and differential geometry, in the language of operator algebras. Thus it should be clear that this is not an introduction to cyclic (co)homology, but only the attempt to single out the basic algebraic facts and techniques of the theory. The reader who wants more motivations should imperatively consult the fundamental article of Alain Connes: Noncommutative differential geometry, I.H.E.S. Publ. Math. vol. 62 (1985), 41-144.

The lectures are organized in two chapters.

The first chapter deals with the intimate relation of cyclic theory to ordinary Hochschild theory, which is at least not surprising by the parallel definition of both theories. There are some important quasi-isomorphisms, proving the equivalence of different approaches to cyclic homology, and spectral sequence techniques are convenient to establish these facts. Thus a comforting paragraph on spectral sequences opens the exposition. Fortunately, I could already take in account the extremely elegant mixed complex approach to cyclic homology of D. Burghelea which streamlines a lot of arguments. The first climax is the fundamental long exact sequence

$$\cdots \to H_n(A) \to HC_n(A) \to HC_{n-2}(A) \to H_{n-1}(A) \to \cdots$$

relating the Hochschild homology groups and the cyclic homology groups (analogously in cohomology), a cornerstone for all structural transmission between both theories. Normalized mixed Hochschild complexes and reduced theory are treated in order to invest conveniently differential ideas: Our operator B becomes a good candidate for a noncommutative outer derivative. Finally, Morita-invariance of Hochschild homology and of cyclic homology are treated, following closely an exposition of K. Igusa.

The second chapter deals with cyclic homology as a typical characteristic zero theory. First, its relation to de Rham cohomology is considered. It comes out that (noncommutative) de Rham cohomology in the sense of M. Karoubi can be embedded in (reduced) cyclic homology. For smooth commutative algebras this can be made more precise by a sort of inverse limit constellation, which is formulated via a decomposition of cyclic homology into ordinary de Rham cohomology:

$$HC_n(A) = \Omega^n/d\Omega^{n-1} \oplus H_{DR}^{n-2}(A) \oplus H_{DR}^{n-4}(A) \oplus \ldots$$

This result of J.L. Loday and D. Quillen has a dual version in continuous cyclic cohomology, due to A. Connes (with $A = C^\infty(X,\mathbb{C})$, the $\mathbb{C}$-algebra of smooth complex-valued functions on a compact manifold X). The final sections of the second chapter deal with cyclic homology as "additive K-theory", in the following sense: For an associative algebra A over a field k of characteristic zero, cyclic homology $HC_*(A)$ is, up to a dimension shift, isomorphic to the space of primitive elements $\mathrm{Prim}\, H_*(gl(A))$ of the Lie algebra homology of $gl(A) = \varinjlim gl_r(A)$. This result should be appreciated in the light of D. Quillen's "multiplicative" version: Rational algebraic K-theory $K_*(A) \otimes \mathbb{Q}$ identifies with $\mathrm{Prim}\, H_*(GL(A),\mathbb{Q})$, the primitive part of (discrete) group homology of $GL(A) = \varinjlim GL_r(A)$. I have to admit two important algebraic omissions: First, I did not treat products (essential in cohomology, which I neglected a bit) and Künneth-formulas (since I got afraid of the costructure invasion). Then, which is perhaps more serious, I did not treat the relations to algebraic K-theory via Chern-characters. On a certain level of arguments this motivates half of the existence of the theory ("create a range for invariants"), but whenever you make the first step towards topology and geometry ...

A few words about the use of spectral sequences in these lectures. We only need them in order to establish some fundamental quasi-isomorphisms, by an approximation argument. This could also be done via explicit and involved calculations. But, since we aim at furnishing the necessary material for further reading, there is no reason to avoid spectral sequence techniques (look at the literature!). At any rate, there is a coherent approach to the basic skeleton of the theory, avoiding all spectral sequence arguments. You begin with I.2.3; then you define cyclic homology of a unital associative k-algebra A as the cyclic homology of the mixed complex $C(A)$ (cf. I.2.4.2). You get all of the material from I.2.4.6 to I.2.4.12. Reduced theory remains unchanged (I.2.5.1 to I.2.5.13). Cyclic cohomology is treated analogously. As to Morita-invariance, the spectral sequence argument in I.2.7.9 is easily replaced by a direct reasoning. de Rham theory (II.1) is already clean. Thus you cover rapidly all of the basic material. It should be pointed out, however, that the equivalence of the different approaches to cyclic (co)homology is an essential part of its handiness.

Finally, I would like to express my thanks to all those who helped me to finish these lectures: first of all, to Daniel Kastler, whose stimulating enthousiasm for the subject and clever support (on many levels) pushed me across this experience. Then, to Joachim Cuntz and Georges Zeller-Meier, who taught me the essentials of the subject, and finally to Philippe Blanchard, Sergio Doplicher, Rudolph Haag and Daniel Testard, whose hospitality and interest at different stages of the work I shall never forget.
This paper was written while the author was guest of the Research Center Bielefeld-Bochum-Stochastics (BiBoS) at the University of Bielefeld. I would like to express my thanks for its kind hospitality. Thanks also to Mrs. Aoyama-Potthoff for the excellent and competent typewriting.

Rome, May 1986 Peter Seibt

Chapter I. Cyclic (co)homology and Hochschild (co)homology.

The fundamental result relating Hochschild and cyclic (co)homology splits in a spectral sequence formulation and a long exact sequence formulation.
Spectral sequence techniques reveal essential, so we begin with an exposition of the relevant material about (a rather special type of) spectral sequences.

I.1 Preliminaries: Spectral Sequences.

I.1.1 Filtered Complexes and Exact Couples.

Definition 1.1.1 Let C be a chain complex (of left R-modules, R a unitary ring). A filtration $(F^pC)_{p\in\mathbb{Z}}$ of C is a family of subcomplexes of C such that $F^{p-1}C \subset F^pC$ for all $p \in \mathbb{Z}$.

Remark 1.1.2 More explicitely, the situation is as follows:

$$\begin{array}{lccccccccc} C: & \cdots & \to & C_{n+1} & \to & C_n & \to & C_{n-1} & \to & \cdots \\ & & & \cup & & \cup & & \cup & & \\ F^pC: & \cdots & \to & F^pC_{n+1} & \to & F^pC_n & \to & F^pC_{n-1} & \to & \cdots \\ & & & \cup & & \cup & & \cup & & \\ F^{p-1}C: & \cdots & \to & F^{p-1}C_{n+1} & \to & F^{p-1}C_n & \to & F^{p-1}C_{n-1} & \to & \end{array}$$

The arrows are the differentials (compatible with the inclusions).

Definition 1.1.3 An exact couple is a quintuple $(D,E,\alpha,\beta,\gamma)$, where D and E are bigraded (at least $\mathbb{Z}$-)modules: $D = (D_{p,q})_{p,q\in\mathbb{Z}}$, $E = (E_{p,q})_{p,q\in\mathbb{Z}}$ and where α, β, γ are homomorphisms of bigraded modules such that the following diagram

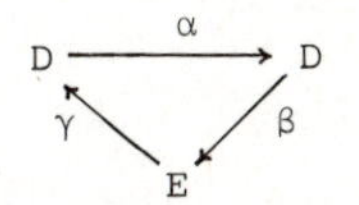

is exact.

Remark 1.1.4 Let (a,a'), (b,b') and (c,c') be the bidegrees of α, β and γ respectively. Then our exact couple consists of an infinity of long exact sequences

$$\longrightarrow E_{p-c,q-c'} \xrightarrow{\gamma} D_{p,q} \xrightarrow{\alpha} D_{p+a,q+a'} \xrightarrow{\beta} E_{p+a+b,q+a'+b'} \longrightarrow$$

$$(\gamma = \gamma_{p-c,q-c'},\ \alpha = \alpha_{p,q},\ \beta = \beta_{p+a,q+a'})$$

Conversely, any such family of long exact sequences defines an exact couple.

Proposition 1.1.5 Every filtration $(F^pC)_{p\in\mathbb{Z}}$ of a chain complex C defines an exact couple

$$\begin{array}{ccc} D & \xrightarrow{\alpha} & D \\ {\scriptstyle\gamma}\nwarrow & & \swarrow{\scriptstyle\beta} \\ & E & \end{array}$$

where α is of bidegree $(1,-1)$

β is of bidegree $(0, 0)$

γ is of bidegree $(-1,0)$.

Proof. For every $p \in \mathbb{Z}$ there is an exact sequence of chain complexes

$$0 \longrightarrow F^{p-1}C \longrightarrow F^pC \longrightarrow F^pC/F^{p-1}C \longrightarrow 0$$

Consider now the long exact homology sequences

$$\longrightarrow H_{p+q}(F^{p-1}C) \xrightarrow{\alpha} H_{p+q}(F^pC) \xrightarrow{\beta} H_{p+q}(F^pC/F^{p-1}C) \xrightarrow{\gamma}$$

$$H_{p+q-1}(F^{p-1}C) \longrightarrow \cdots$$

(The decomposition $n = p+q$ of the grading index n relative to the filtration index p will reveal pretty when dealing with spectral sequences of double complexes)

α is induced by inclusion of chain complexes,

β is induced by natural surjection of chain complexes,

γ is the connection homomorphism.

Define $D_{p,q} = H_{p+q}(F^pC)$

$$E_{p,q} = H_{p+q}(F^pC/F^{p-1}C), \qquad p,\ q \in \mathbb{Z}.$$

The long exact homology sequences can be rewritten as

$$\cdots \longrightarrow D_{p-1,q+1} \xrightarrow{\alpha} D_{p,q} \xrightarrow{\beta} E_{p,q} \xrightarrow{\gamma} D_{p-1,q} \longrightarrow \cdots$$

which establishes our exact couple.
α is of bidegree $(1,-1)$, β is of bidegree $(0,0)$, γ is of bidegree $(-1,0)$, as desired.

I.1.2 The Spectral Sequence associated with an Exact Couple.

Construction 1.2.1 The derived exact couple $(D^2,E^2,\alpha^2,\beta^2,\gamma^2)$ of an exact couple $(D,E,\alpha,\beta,\gamma) = (D^1,E^1,\alpha^1,\beta^1,\gamma^1)$.

Consider an exact couple

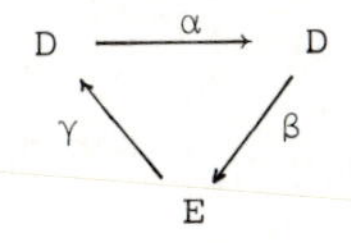

with α, β, γ of bidegrees $(1,-1)$, $(0,0)$, $(-1,0)$ respectively.

We shall construct an exact couple

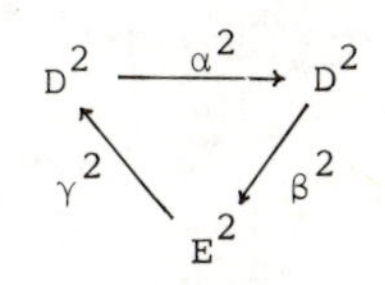

the derived exact couple of $(D,E,\alpha,\beta,\gamma)$ such that
α^2 is of bidegree $(1,-1)$ (as α)
β^2 is of bidegree $(-1,1)$
γ^2 is of bidegree $(-1,0)$ (as γ).

(a) Definition of E^2:
Consider $d^1 : E \to E$ given by $d^1 = \beta\gamma$.

Since $d^1_{p,q} : E_{p,q} \xrightarrow{\gamma} D_{p-1,q} \xrightarrow{\beta} E_{p-1,q}$, we have : d^1 is of bidegree $(-1,0)$, and $d^1d^1 = 0$ (since $\gamma\beta = 0$).

$E^2 = H(E,d^1) = \text{Ker } d^1/\text{Im } d^1$, i.e.

$E^2_{p,q} = \text{Ker } d^1_{p,q}/\text{Im } d^1_{p+1,q}$ for $p,q \in \mathbb{Z}$.

(b) Definition of D^2:

$D^2 = \text{Im } \alpha$, i.e. $D^2_{p,q} = \text{Im } \alpha_{p-1,q+1} \subset D_{p,q}$.

(c) Definition of α^2, β^2 and γ^2:

$$\begin{array}{ccc} D^2 & \xrightarrow{\alpha^2} & D^2 \\ & \gamma^2 \nwarrow \quad \swarrow \beta^2 & \\ & E^2 & \end{array}$$

$\alpha^2 = \alpha | D^2$ (of (bidegree (1,-1), as α)

$\beta^2 : D^2 \to E^2$ is defined as follows:

$$\beta^2_{p,q}(\alpha_{p-1,q+1}(x_{p-1,q+1})) = [\beta_{p-1,q+1}(x_{p-1,q+1})] \in E^2_{p-1,q+1}$$

where $[\cdots]$ means residue class.

β^2 is well-defined:

* $\beta_{p-1,q+1}(x_{p-1,q+1}) \in \text{Ker } d^1_{p-1,q+1}$ since

$d^1_{p-1,q+1} = \beta_{p-2,q+1}\gamma_{p-1,q+1}$

** $\alpha_{p-1,q+1}(x_{p-1,q+1}) = \alpha_{p-1,q+1}(y_{p-1,q+1})$ implies

$x_{p-1,q+1} - y_{p-1,q+1} \in \text{Ker } \alpha_{p-1,q+1} = \text{Im } \gamma_{p,q+1}$, hence

$\beta_{p-1,q+1}(x_{p-1,q+1}) \equiv \beta_{p-1,q+1}(y_{p-1,q+1}) \text{mod Im } d^1_{p,q+1}$

β^2 is of bidegree (-1,1).

$\gamma^2 : E^2 \to D^2$ is defined by γ:

$\gamma^2_{p,q}[z_{p,q}] = \gamma_{p,q}(z_{p,q}) \in D^2_{p-1,q}$

γ^2 is well-defined:

* $z_{p,q} \in \text{Ker } d^1_{p,q}$, $d^1_{p,q} = \beta_{p-1,q}\gamma_{p,q}$, hence

$\gamma_{p,q}(z_{p,q}) \in \text{Ker } \beta_{p-1,q} = \text{Im } \alpha_{p-2,q+1} = D^2_{p-1,q}$

** $z_{p,q} \in \text{Im } d^1_{p+1,q}$, $z_{p,q} = \beta_{p,q}\gamma_{p+1,q}(u_{p+1,q})$ then

$\gamma_{p,q}(z_{p,q}) = \gamma_{p,q}\beta_{p,q}\gamma_{p+1,q}(u_{p+1,q}) = 0$

γ^2 has bidegree (-1,0).

(d) Verification of exactness:

(i) Ker β^2 = Im α^2
(ii) Ker γ^2 = Im β^2
(iii) Ker α^2 = Im γ^2

The inclusions image $\subset$ Kernel are trivial, since α^2, β^2 and γ^2 are induced by α, β and γ. We have to show the reverse inclusions. For notational convenience we shall suppress indices

(i) Ker $\beta^2 \subset$ Im α^2:

$x \in \text{Ker } \beta^2 \subset D^2 = \text{Im } \alpha$ can be written as $x = \alpha(u)$, and $\beta^2(x) = [\beta u] = 0$, i.e. $\beta(u) \in \text{Im } d^1$. There is $w \in E$: $\beta(u) = \beta\gamma(w)$, hence $u - \gamma(w) \in \text{Ker } \beta = \text{Im } \alpha = D^2$. But $\alpha^2(u-\gamma(w)) = \alpha(u) - \alpha\gamma(w) = \alpha(u) = x$, i.e. $x \in \text{Im } \alpha^2$.

(ii) Ker $\gamma^2 \subset$ Im β^2:

Consider $x = [z] \in E^2$ such that $\gamma^2(x) = \gamma(z) = 0$. $z \in \text{Ker } \gamma = \text{Im } \beta$; write $z = \beta(w)$. Then $x = [z] = [\beta(w)] = \beta^2(\alpha(w)) \in \text{Im } \beta^2$.

(iii) Ker $\alpha^2 \subset$ Im γ^2:

For $x \in D^2 = \text{Im } \alpha$ such that $\alpha^2(x) = \alpha(x) = 0$ we have: $x \in \text{Im } \gamma = \text{Ker } \alpha$, i.e. there is $y \in E$ with $x = \gamma(y)$. We will have $x = \gamma^2[y] \in \text{Im } \gamma^2$ provided that $y \in \text{Ker } d^1 = \text{Ker } \beta\gamma$. But $x \in \text{Im } \alpha = \text{Ker } \beta$, hence $\beta\gamma(y) = \beta(x) = 0$.

<u>Example 1.2.2</u> Let C be a filtered chain complex, and let $(D,E,\alpha,\beta,\gamma)$ be the exact couple associated with the filtration $(F^pC)_{p\in\mathbb{Z}}$ of C. We shall determine $E^2_{p,q} = \text{Ker } d^1_{p,q}/\text{Im } d^1_{p+1,q}$. Consider

$$\begin{array}{ccccccc} H_{p+q}(F^{p-1}C) & \longrightarrow & H_{p+q}(F^pC) & \longrightarrow & H_{p+q}(F^pC/F^{p-1}C) & \longrightarrow & H_{p+q-1}(F^{p-1}C) \\ \| & & \| & & \| & & \| \\ D_{p-1,q+1} & \xrightarrow{\alpha} & D_{p,q} & \xrightarrow{\beta} & E_{p,q} & \xrightarrow{\gamma} & D_{p-1,q} \end{array}$$

where $\gamma_{p,q} : E_{p,q} \to D_{p-1,q}$ is the connecting homomorphism (cf. 1.1.5). We have exact sequences of chain complexes

$$0 \to F^{p-1}C/F^{p-2}C \to F^pC/F^{p-2}C \to F^pC/F^{p-1}C \to 0$$

giving rise to connecting homomorphisms

$$\partial_{p+q} : H_{p+q}(F^pC/F^{p-1}C) \to H_{p+q-1}(F^{p-1}C/F^{p-2}C).$$

Let us show that $d^1_{p,q} : E_{p,q} \to E_{p-1,q}$ identifies with ∂_{p+q}. To see this consider the commutative diagram of homomophisms of chain complexes

$$\begin{array}{ccccccccc}
0 & \longrightarrow & F^{p-1}C & \longrightarrow & F^pC & \longrightarrow & F^pC/F^{p-1}C & \to & 0 \\
 & & \downarrow & & \downarrow & & \| & & \\
0 & \to & F^{p-1}C/F^{p-2}C & \longrightarrow & F^pC/F^{p-2}C & \longrightarrow & F^pC/F^{p-1}C & \to & 0
\end{array}$$

with exact rows. We obtain the following commutative diagram relating the two long exact homology sequences in question:

$$\begin{array}{ccccccc}
H_{p+q}(F^{p-1}C) & \xrightarrow{\alpha} & H_{p+q}(F^pC) & \xrightarrow{\beta} & H_{p+q}(F^pC/F^{p-1}C) & \xrightarrow{\gamma} & H_{p+q-1}(F^{p-1}C) \\
\downarrow\beta & & \downarrow & & \| & & \downarrow\beta \\
H_{p+q}(F^{p-1}C/F^{p-2}C) & \longrightarrow & H_{p+q}(F^pC/F^{p-2}C) & \longrightarrow & H_{p+q}(F^pC/F^{p-1}C) & \xrightarrow{\partial} & H_{p+q-1}(F^{p-1}C/F^{p-2}C)
\end{array}$$

This yields $\partial_{p+q} = \beta_{p-1,q}\gamma_{p,q} = d^1_{p,q}$, as desired.

<u>Proposition 1.2.3</u> Let $(D,E,\alpha,\beta,\gamma)$ be an exact couple, with α, β, γ of bidegrees $(1,-1)$, $(0,0)$, $(-1,0)$ respectively. Let $(D^r,E^r,\alpha^r,\beta^r,\gamma^r)$, $r \geq 2$, be the (r-1)th derived couple of $(D,E,\alpha,\beta,\gamma)$ (iterate the construction 1.2.1). Then we have

(1) α^r has bidegree $(1,-1)$, β^r has bidegree $(1-r,r-1)$, γ^r has bidegree $(-1,0)$.

(2) $d^r = \beta^r\gamma^r$ has bidegree $(-r,r-1)$.

(3) $E^{r+1}_{p,q} = \text{Ker } d^r_{p,q}/\text{Im } d^r_{p+r,q-r+1}$.

<u>Proof</u>. Trivial. Observe that the construction 1.2.1 goes through with α, β and γ of arbitrary bidegrees. In the situation $(1,-1)$, (b,b'), $(-1,0)$ β^2 will have bidegree equal to $(b-1,b'+1)$.

<u>Definition 1.2.4</u> Let $(D,E,\alpha,\beta,\gamma) = (D^1,E^1,\alpha^1,\beta^1,\gamma^1)$ be an exact couple. The sequence $(E^r,d^r)_{r\geq 1}$ is called the <u>spectral sequence</u> associated with the exact couple.

Every filtered chain complex C thus gives rise to a spectral sequence.

<u>Remark 1.2.5</u> There is a chain of submodules of the bigraded module E^2

$$0 \subset B^2 \subset B^3 \subset \cdots \subset B^r \subset \cdots \subset Z^r \subset \cdots \subset Z^3 \subset Z^2 \subset E^2$$

such that $Z^r/B^r \simeq E^{r+1}$, $r \geq 2$. This is easily seen by induction on r.

$r = 2$: $Z^2 = \operatorname{Ker} d^2 \subset E^2$, $B^2 = \operatorname{Im} d^2 \subset E^2$, $Z^2/B^2 = E^3$

$r \mapsto r+1$: We have $E^{r+2} = \operatorname{Ker} d^{r+1}/\operatorname{Im} d^{r+1}$, and $\operatorname{Im} d^{r+1} \subset \operatorname{Ker} d^{r+1} \subset E^{r+1} \simeq Z^r/B^r$,

with $0 \subset B^2 \subset \cdots \subset B^r \subset Z^r \subset \cdots \subset Z^2 \subset E^2$. Let $\nu_r : Z^r \to E^{r+1}$ be the epimorphism with $B^r = \operatorname{Ker} \nu_r$. Define $Z^{r+1} = \nu_r^{-1} \operatorname{Ker} d^{r+1}$, $B^{r+1} = \nu_r^{-1} \operatorname{Im} d^{r+1}$. Then $B^r \subset B^{r+1} \subset Z^{r+1} \subset Z^r$ and $Z^{r+1}/B^{r+1} \simeq E^{r+2}$.

Notation: $Z^r_{p,q} = Z^r \cap E^2_{p,q}$, $B^r_{p,q} = B^r \cap E^2_{p,q}$, $Z^\infty_{p,q} = \bigcap_{r\geq 2} Z^r_{p,q}$,

$B^\infty_{p,q} = \bigcup_{r\geq 2} B^r_{p,q}$, $E^\infty_{p,q} = Z^\infty_{p,q}/B^\infty_{p,q}$

Definition 1.2.6 The bigraded module $E^\infty = (E^\infty_{p,q})_{p,q\in\mathbb{Z}}$ will be called the limit term of the spectral sequence $(E^r,d^r)_{r\geq 1}$.

Note that if you define a spectral sequence $(E^r,d^r)_{r\geq 1}$ as a sequence of differential modules such that $H(E^r,d^r) = E^{r+1}$, $r \geq 1$, then 1.2.5 and 1.2.6 make sense (forget the bigradings).

I.1.3 Convergence of a Spectral Sequence.

For our purposes it suffices to treat (locally) finite convergence.

Remark 1.3.1 Let $H = (H_n)_{n\in\mathbb{Z}}$ be a graded module. A filtration $(F^pH)_{p\in\mathbb{Z}}$ of H will always be understood as being given by a chain of graded submodules, i.e. we may write $F^pH = (F^pH_n)_{n\in\mathbb{Z}}$, $p \in \mathbb{Z}$.

Definition 1.3.2 In the situation of 1.3.1 the filtration is called bounded, whenever the following holds: For every $n \in \mathbb{Z}$ there exists $s = s(n)$, $t = t(n)$ such that $F^sH_n = 0$, $F^tH_n = H_n$, i.e. for every n the filtration of H_n is given by a finite chain $0 = F^sH_n \subset F^{s+1}H_n \subset \cdots \subset F^{t-1}H_n \subset F^tH_n = H_n$ (the length of which generally depends on n).

Note that a chain complex C is a graded module, hence boundedness of a filtration $(F^pC)_{p\in\mathbb{Z}}$ of C (by a chain of subcomplexes) is thus defined.

Definition 1.3.3 Let $H = (H_n)_{n\in\mathbb{Z}}$ be a graded module, $(E^r,d^r)_{r\geq 1}$ a spectral sequence associated with an exact couple.

$E^2_{p,q} \underset{p}{\Rightarrow} H_n$ ("the spectral sequence converges to H") whenever there is a bounded filtration $(\phi^pH)_{p\in\mathbb{Z}}$ of the graded module H such that

$E^\infty_{p,q} \simeq \phi^p H_n/\phi^{p-1}H_n$ for every $(p,q) \in \mathbb{Z}^2$.

Notational convention: $n = p + q$!

Remark 1.3.4 When speaking of modules, we never explicitely specify the ring of scalars, say R. The technical problems about spectral sequences (as far as we treat them) don't involve any knowledge about R. A statement as in 1.3.3 is thus to be interpreted "over a common R" (which is at least $\mathbb{Z}$, the integers).

Proposition 1.3.5 Let $(F^pC)_{p\in\mathbb{Z}}$ be a bounded filtration of a chain complex C, and let $(E^r,d^r)_{r\geq 1}$ be the spectral sequence associated with it. Then we have:

(1) For every (p,q) there is an $r_o = r_o(p,q)$ such that $E^\infty_{p,q} = E^r_{p,q}$ for $r \geq r_o$.

(2) $E^2_{p,q} \underset{p}{\Rightarrow} H_n(C)$.

Proof.

(1)(i) Fix $n \in \mathbb{Z}$, and consider the "line" $n = p + q$ in the lattice of indices.

We have $F^{p-1}C_n = F^pC_n$ whenever $p < s(n)$, $t(n) < p$ (boundedness of the filtration of C). Thus $E_{p,q} = H_{p+q}(F^pC/F^{p-1}C) = 0$ for $p < s(n)$, $t(n) < p$, $q = n - q$, i.e. $E^r_{p,q} = 0$ for $r \geq 1$ and this index-constellation.

(ii) $d^r : E^r \to E^r$ has bidegree $(-r,r-1)$, i.e. $d^r_{p,q}(E^r_{p,q}) \subset E^r_{p-r,q+r-1}$. For fixed $(p,q) \in \mathbb{Z}^2$ we thus have by (i): $E^r_{p-r,q+r-1} = 0 = E^r_{p+r,q-r+1}$ whenever $r \geq r_o(p,q)$ which means that $E^r_{p,q} = \operatorname{Ker} d^r_{p,q}$ and $E^{r+1}_{p,q} = \operatorname{Ker} d^r_{p,q}/\operatorname{Im} d^r_{p+r,q-r+1} = E^r_{p,q}$ for $r \geq r_o(p,q)$.

This yields $E^\infty_{p,q} = E^r_{p,q}$ for sufficiently large r.

(2)(i) Definition of a bounded filtration $(\phi^pH(C))_{p\in\mathbb{Z}}$ on $H(C)$. We set $\phi^pH_n(C) = \operatorname{Im}(H_n(F^pC) \to H_n(C))$. The finite chain $0 = F^sC_n \subset F^{s+1}C_n \subset\cdots\subset F^tC_n = C_n$ yields immediately a finite chain $0 = \phi^sH_n(C) \subset \phi^{s+1}H_n(C) \subset\cdots\subset \phi^tH_n(C) = H_n(C)$ (i.e. the bounds $s(n)$ and $t(n)$ are the same for both filtrations).

(ii) Look now at the (r-1)th derived couple:

$$E^r_{p+r-1,q-r+2} \xrightarrow{\gamma^r} D^r_{p+r-2,q-r+2} \xrightarrow{\alpha^r} D^r_{p+r-1,q-r+1} \xrightarrow{\beta^r} E^r_{p,q} \xrightarrow{\gamma^r} D^r_{p-1,q} \to \cdots$$

We will show that for r sufficiently large (depending on p,q) we have:

$$E^r_{p+r-1,q-r+2} = 0 = D^r_{p-1,q}$$

$$D^r_{p+r-2,q-r+2} = \phi^{p-1}H_{p+q}(\mathcal{C})$$

$$D^r_{p+r-1,q-r+1} = \phi^{p}H_{p+q}(\mathcal{C})$$

(which will prove the assertion (2) of our proposition).

The vanishing of the E-term for large r has already been established. We only have to treat the D-terms.

$D^r = \underbrace{\alpha \circ \alpha \circ \cdots \circ \alpha}_{r-1 \text{ times}} D$, and α has bidegree $(1,-1)$.

Recall that α is induced by one-step-inclusion of the filtration of $\mathcal{C}$; hence

$$D^r_{p+r-1,q-r+1} = \underbrace{\alpha \circ \alpha \circ \cdots \circ \alpha}_{r-1 \text{ times}} D_{p,q} = \operatorname{Im}(H_{p+q}(F^p\mathcal{C}) \to H_{p+q}(F^{p+r-1}\mathcal{C})).$$

For fixed $(p,q) \in \mathbb{Z}^2$, $n = p + q$, we have $F^{p+r-1}C_n = C_n$ whenever r is sufficiently large. This yields

$$D^r_{p+r-1,q-r+1} = \operatorname{Im}(H_{p+q}(F^p\mathcal{C}) \to H_{p+q}(\mathcal{C})) = \phi^p H_{p+q}(\mathcal{C})$$

Analogously we obtain $D^r_{p+r-2,q-r+2} = \phi^{p-1}H_{p+q}(\mathcal{C})$ for large r.

Finally, $D^r_{p-1,q} = 0$ for large r, since

$$D^r_{p-1,q} = \underbrace{\alpha \circ \alpha \circ \cdots \circ \alpha}_{r-1 \text{ times}} D_{p-r,q+r-1} = \alpha \circ \alpha \circ \cdots \circ \alpha H_{p+q-1}(F^{p-r}\mathcal{C})$$

and $H_{p+q-1}(F^{p-r}\mathcal{C}) = 0$ whenever $p-r < s = s(p+q-1)$.

Remark 1.3.6 Let $f : \mathcal{C} \to \mathcal{D}$ be a homomorphism of filtered chain complexes (compatible with the filtrations, gradings and the differentials). f induces for every $r \geq 1$ a homomorphism (of bidegree $(0,0)$)

$$f^r : E^r(\mathcal{C}) \to E^r(\mathcal{D}).$$

To see this, consider the commutative diagram

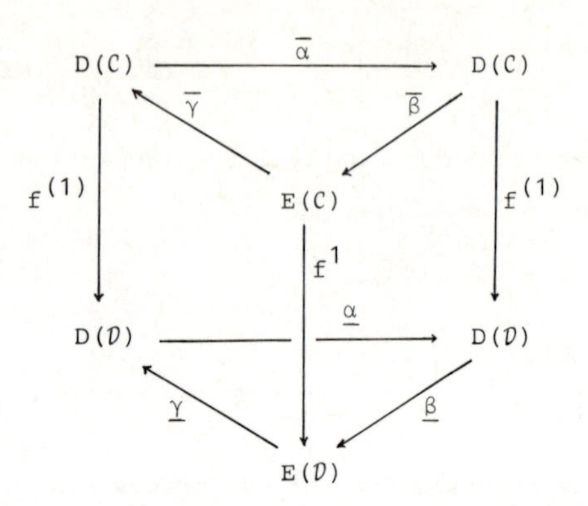

where $f^{(1)}$ and f^1 are induced by f in homology.

It is immediate that we obtain thus a commutative diagram of the same kind between the derived couples (where now $f^{(2)}$ and f^2 are induced by $f^{(1)}$ and f^1, hence by f); iteration gives all we want.

The functoriality of E^r (arrows of filtered chain complexes → arrows of bigraded differential modules) follows from the functoriality of homology.

Theorem 1.3.7 (Approximation theorem)
Let $f : \mathcal{C} \to \mathcal{D}$ be a homomorphism of filtered chain complexes, both with bounded filtrations. If there exists an $r \geq 1$ such that $f^r : E^r(\mathcal{C}) \to E^r(\mathcal{D})$ is an isomorphism, then f is a quasi-isomorphism (i.e. $H(f): H(\mathcal{C}) \to H(\mathcal{D})$ is an isomorphism).

Proof.
(a) Recall the proof of 1.3.5 (2)(ii)

$$D^r_{p+r-2,q-r+2} = \mathrm{Im}(H_{p+q}(F^{p-1}\mathcal{C}) \to H_{p+q}(F^{p+r-2}\mathcal{C}))$$

$$D^r_{p+r-1,q-r+1} = \mathrm{Im}(H_{p+q}(F^{p}\mathcal{C}) \to H_{p+q}(F^{p+r-1}\mathcal{C}))$$

$$E^r_{p,q} = \mathrm{Im}(H_{p+q}(F^{p}\mathcal{C}) \to H_{p+q}(F^{p+r-1}\mathcal{C}/F^{p-1}\mathcal{C}))$$

(which follows from the definition of the derived exact couples), where the arrows

$$D^r_{p+r-2,q-r+2} \xrightarrow{\alpha^r} D^r_{p+r-1,q-r+1} \xrightarrow{\beta^r} E^r_{p,q}$$

are the obvious ones.
For large r (depending on p,q) we obtained the exact sequence

$$0 \to \phi^{p-1}H_{p+q}(C) \to \phi^{p}H_{p+q}(C) \to E^{\infty}_{p,q} \to 0$$

where $E^{\infty}_{p,q} = \mathrm{Im}(H_{p+q}(F^{p}C) \to H_{p+q}(C/F^{p-1}C))$. The isomorphism $E^{\infty}_{p,q} \simeq \phi^{p}H_{p+q}/\phi^{p-1}H_{p+q}$ is thus induced by the commutative diagram (with exact row)

$$\begin{array}{ccccc} & \nearrow & H_{p+q}(F^{p}C) & \searrow & \\ & & \downarrow & & \\ H_{p+q}(F^{p-1}C) & \longrightarrow & H_{p+q}(C) & \longrightarrow & H_{p+q}(C/F^{p-1}C) \end{array}$$

(b)(i) We have $f^{\infty} : E^{\infty}(C) \to E^{\infty}(\mathcal{D})$ with $f^{\infty}_{p,q}: E^{\infty}_{p,q}(C) \to E^{\infty}_{p,q}(\mathcal{D})$ induced by f (actually $f^{r}_{p,q} : E^{r}_{p,q}(C) \to E^{r}_{p,q}(\mathcal{D})$, $r \geq r_o(p,q)$). On the other hand, $H(f) : H(C) \to H(\mathcal{D})$ respects the filtrations $(\phi^{p}H(C))_{p\in\mathbb{Z}}$ and $(\psi^{p}H(\mathcal{D}))_{p\in\mathbb{Z}}$ (and the gradings), thus induces

$gr(H(f)): gr_{\phi}H(C) \to gr_{\psi}H(\mathcal{D})$ with

$gr(H(f))_{p,q}: \phi^{p}H_{p+q}(C)/\phi^{p-1}H_{p+q}(C) \to \psi^{p}H_{p+q}(\mathcal{D})/\psi^{p-1}H_{p+q}(\mathcal{D})$

(ii) By virtue of part (a) of our proof we obtain a commutative diagram

$$\begin{array}{ccc} E^{\infty}_{p,q}(C) & \xrightarrow{\quad f^{\infty}_{p,q} \quad} & E^{\infty}_{p,q}(\mathcal{D}) \\ \| & & \| \\ \phi^{p}H_{p+q}(C)/\phi^{p-1}H_{p+q}(C) & \xrightarrow{gr(H(f))_{p,q}} & \psi^{p}H_{p,q}(\mathcal{D})/\psi^{p-1}H_{p+q}(\mathcal{D}) \end{array}$$

("everything is induced by f in homology").

Conclusion: If f^{r} is an isomorphism, then f^{s} is an isomorphism for $s \geq r$, hence f^{∞} is an isomorphism, i.e. $gr(H(f)): gr_{\phi}H(C) \to gr_{\psi}H(\mathcal{D})$ is an isomorphism.
We have to show that $H(f): H(C) \to H(\mathcal{D})$ is an isomorphism.

(iii) The boundedness of the filtrations on $H(C)$ and $H(\mathcal{D})$ now give the desired conclusion:

$H(f)$ is injective since $gr(H(f))$ is injective.

$H(f)$ is surjective since $gr(H(f))$ is surjective.

Let us write $h = H(f)$, $\bar{h} = gr(H(f))$.

Choose first $x \in \phi^p H_{p+q}(C)$ such that $h(x) = 0$. Then $\bar{h}_{p,q}(\bar{x}) = 0$ in $\psi^p H_{p+q}(\mathcal{D})/\psi^{p-1} H_{p+q}(\mathcal{D})$, i.e. $x \in \phi^{p-1} H_{p+q}(C)$ ($\bar{h}$ being injective).

Now iterate: $x \in \phi^{p-K} H_{p+q}(C)$ for all $K \geq 0$, hence $x = 0$ (boundedness of the ϕ-filtration).

Take now $y \in \psi^p H_{p+q}(\mathcal{D})$. We have to find $x \in H_{p+q}(C)$ such that $h(x) = y$.

By the surjectivity of $\bar{h}$ there is $x_o \in \phi^p H_{p+q}(C)$ with $\bar{h}_{p,q}(\bar{x}_o) = \bar{y}$, i.e. $y \equiv h(x_o) \bmod \psi^{p-1} H_{p+q}(\mathcal{D})$. Continue with $y_1 = y - h(x_o) \in \psi^{p-1}\cdot H_{p+q}(\mathcal{D})$. You will obtain $x_1 \in \phi^{p-1} H_{p+q}(C)$ such that $y_2 = y_1 - h(x_1)$ is contained in $\psi^{p-2} H_{p+q}(\mathcal{D})$.

Iterating, you obtain $x_K \in \phi^{p-K} H_{p+q}(C)$ such that

$y_K \equiv h(x_K) \bmod \psi^{p-K-1} H_{p+q}(\mathcal{D})$.

The boundedness of the ψ-filtration yields some $s \geq 0$ with $y_{s+1} = 0$.

Then: $y = h(\sum_{i=0}^{s} x_i) = h(x_o) + h(x_1) + \cdots + h(x_s)$.

I.1.4 Double Complexes and their Spectral Sequences.

Definition 1.4.1 A double complex is a triple (M, d', d''), when $M = (M_{p,q})_{p,q \in \mathbb{Z}}$ is a bigraded module and where $d', d'' : M \to M$ are homomorphisms such that

(i) d' has bidegree $(-1,0)$, $d' \circ d' = 0$

(ii) d'' has bidegree $(0,-1)$, $d'' \circ d'' = 0$

(iii) $d' \circ d'' + d'' \circ d' = 0$

Remark 1.4.2

$$d'_{p,q} : M_{p,q} \to M_{p-1,q}$$

$$d''_{p,q} : M_{p,q} \to M_{p,q-1}$$

Every row $M_{p,*}$ gives rise to a chain complex $(M_{p,*}, d''_{p,*})$, and every column $M_{*,q}$ gives rise to a chain complex $(M_{*,q}, d'_{*,q})$. (M,d') and (M,d'') are chain complexes (forget one of the differentials).

Definition 1.4.3 Let (M, d', d'') be a double complex. $(\mathrm{Tot}(M), d)$, the

<u>total</u> (chain) <u>complex</u> of (M,d',d'') is given by

(i) $\mathrm{Tot}(M)_n = \bigoplus_{p+q=n} M_{p,q}, \quad n \in \mathbb{Z}$

(ii) $d_n : \mathrm{Tot}(M)_n \to \mathrm{Tot}(M)_{n-1}$

$$d_n/M_{p,q} = d'_{p,q} + d''_{p,q}$$

<u>Remark 1.4.4</u> $d_{n-1} \circ d_n = 0$ by virtue of $d'_{p-1,q} \circ d'_{p,q} = 0$, $d''_{p,q-1} \circ d''_{p,q} = 0$, $d''_{p-1,q} \circ d'_{p,q} + d'_{p,q-1} \circ d''_{p,q} = 0$

<u>Example 1.4.5</u> Let $M = (M_{p,q})_{p,q \in \mathbb{Z}}$ be a bigraded module, d', d'': $M \to M$ two differentials of bidegree $(-1,0)$ and $(0,-1)$ respectively which commute.

Define $\delta''_{p,q} = (-1)^p d''_{p,q}$. Then (M,d',δ'') is a double complex.

<u>Definition 1.4.6</u> Let (M,d',d'') be a double complex. The first and the second filtration of $\mathrm{Tot}(M)$:

$$({}^{\mathrm{I}}F^p\mathrm{Tot}(M))_n = \bigoplus_{i \le p} M_{i,n-i}$$

$$({}^{\mathrm{II}}F^p\mathrm{Tot}(M))_n = \bigoplus_{j \le p} M_{n-j,j} \quad \text{(Attention: } p \text{ restricts the second index!)}$$

<u>Remark 1.4.7</u> For every $p \in \mathbb{Z}$, ${}^{\mathrm{I}}F^p\mathrm{Tot}(M)$ is a subcomplex of $\mathrm{Tot}(M)$. Let $M_{i,n-i}$ be a component of $({}^{\mathrm{I}}F^p\mathrm{Tot}(M))_n$, i.e. $i \le p$. Then

$d_n M_{i,n-i} = d'_{i,n-i} M_{i,n-i} + d''_{i,n-i} M_{i,n-i} \subset M_{i-1,n-i} \oplus M_{i,n-i-1}$ i.e.

$d_n M_{i,n-i} \subset ({}^{\mathrm{I}}F^p\mathrm{Tot}(M))_{n-1}$.

Anologously: ${}^{\mathrm{II}}F^p\mathrm{Tot}(M)$ is a subcomplex of $\mathrm{Tot}(M)$.

<u>Lemma 1.4.8</u> Let (M,d',d'') be a double complex. $({}^{\mathrm{I}}F^p\mathrm{Tot}(M))_{p \in \mathbb{Z}}$ and $({}^{\mathrm{II}}F^p\mathrm{Tot}(M))_{p \in \mathbb{Z}}$ are bounded filtrations if and only if for every $n \in \mathbb{Z}$ there is only a finite number of $(p,q) \in \mathbb{Z}^2$ such that $p+q = n$ and $M_{p,q} \neq 0$.

<u>Proof</u>. Trivial.

<u>Remark-Corollary 1.4.9</u> Let (M,d',d'') be a double complex of the first or of the third quadrant (obvious vanishing conditions on the

$M_{p,q}$), and let $({}^{I}E^{r})_{r\geq 1}$ and $({}^{II}E^{r})_{r\geq 1}$ be the spectral sequences determined by the first and second filtration on Tot(M). Then we have

(1) ${}^{I}E^{\infty}_{p,q} = {}^{I}E^{r}_{p,q}$ for r sufficiently large (depending on p,q). Analogous statement for the second spectral sequence.

(2) ${}^{I}E^{2}_{p,q} \underset{p}{\Rightarrow} H_n(\mathrm{Tot}(M))$

${}^{II}E^{2}_{p,q} \underset{p}{\Rightarrow} H_n(\mathrm{Tot}(M))$

Proof. Immediate by 1.3.5 and 1.4.8.

Remark 1.4.10 We want to determine ${}^{I}E^{2}_{p,q}$ (and ${}^{II}E^{2}_{p,q}$). Let us attack the first filtration. Write $T = \mathrm{Tot}(M)$, and drop the upper-I-index. We have $E_{p,q} = H_{p+q}(F^pT/F^{p-1}T)$.

But $(F^pT/F^{p-1}T)_{p+q} = \bigoplus_{i\leq p} M_{i,p+q-i} / \bigoplus_{i\leq p-1} M_{i,p+q-i} = M_{p,q}$.

Thus $F^pT/F^{p-1}T$ is the pth row $M_{p,*}$ of M, with differential $d'' = d''_{p,*}$ (actually, the differential of $F^pT/F^{p-1}T$ is induced by $d = d'+d''$, but d' goes vertically, i.e. $d'F^pT \subset F^{p-1}T$, hence d' induces the zero-homomorphism on $F^pT/F^{p-1}T$).

Finally: $E_{p,q} = \mathrm{Ker}\, d''_{p,q}/\mathrm{Im}\, d''_{p,q+1} = H''_{p,q}(M)$

(where H'' denotes the homology of the chain complex (M,d'')).

For every $q \in \mathbb{Z}$ we get a chain complex

$$\cdots \longleftarrow H''_{p-1,q}(M) \longleftarrow H''_{p,q}(M) \longleftarrow H''_{p+1,q}(M) \longleftarrow \cdots$$

with differential $\bar{d}'$ induced by d':

$$\bar{d}'_{p,q}[z_{p,q}] = [d'_{p,q} z_{p,q}]$$

(this is well-defined, since $d' \circ d'' = -d'' \circ d'$).

When passing to homology, we obtain a bigraded module $(H'_pH''_{p,q}(M))_{p,q\in\mathbb{Z}}$ associated with the double complex (M,d',d'').

Proposition 1.4.11 In the situation 1.4.10 we have ${}^{I}E^{2}_{p,q} = H'_pH''_{p,q}(M)$.

Proof. We already know that $E_{p,q} = H''_{p,q}(M)$, and that $E^2_{p,q} =$

$\mathrm{Ker}\ d^1_{p,q}/\mathrm{Im}\ d^1_{p+1,q}$, where $d^1_{p,q} : E_{p,q} \to E_{p-1,q}$ identifies with the connecting homomorphism

$\partial_{p,q}: H''_{p,q}(M) \to H''_{p-1,q}(M)$ of the long exact homology sequence associated with

$$0 \longrightarrow F^{p-1}T/F^{p-2}T \longrightarrow F^pT/F^{p-2}T \longrightarrow F^pT/F^{p-1}T \longrightarrow 0$$

The explicit description of $\partial_{p,q}$ is as follows: Look at our exact sequence of chain complexes in degree $n = p+q$ and $n-1 = p+q-1$:

$$\begin{array}{ccccccc} & & M_{p,q} \oplus M_{p-1,q+1} & \xrightarrow{\pi} & M_{p,q} & \longrightarrow & 0 \\ & & \downarrow{\scriptstyle d'+d''} & & & & \\ 0 \longrightarrow M_{p-1,q} & \xrightarrow{i} & M_{p-1,q} \oplus M_{p,q-1} & & & & \end{array}$$

Let $z \in M_{p,q}$ represent $x \in H''_{p,q}(M)$. We have to identify $\partial_{p,q}(x)$ $\in H''_{p-1,q}(M)$. By definition of the connecting homomorphism you have to choose $(z,0) \in \pi^{-1}(z)$, and to pass to $i^{-1}d(z,0) = d'(z)$ $(d''(z) = 0!)$

$\partial_{p,q}(x) = [d'z] \in H''_{p-1,q}(M)$.

Thus $d^1_{p,q} = \partial_{p,q} = \overline{d}'$ (1.4.10), i.e.

$E^2_{p,q} = \mathrm{Ker}\ d^1_{p,q}/\mathrm{Im}\ d^1_{p+1,q} = H'_pH''_{p,q}(M)$, as desired.

<u>Corollary 1.4.12</u> For a double complex (M,d',d'') we have ${}^{II}E^2_{p,q} =$ $H''_pH'_{q,p}(M)$ (Attention: look at the subscripts!)

<u>Proof</u>. Reduction to the first filtration case. Define the transposed double complex (M^t,Δ',Δ'') by:

$M^t_{p,q} = M_{q,p}$, $\Delta'_{p,q} = d''_{q,p}$, $\Delta''_{p,q} = d'_{q,p}$,

Then $\mathrm{Tot}(M^t) = \mathrm{Tot}(M)$, with the same differential.

We have thus

$$\begin{array}{ccc} {}^{I}E^{t\,2}_{p,q} & = & H'_pH''_{p,q}(M^t) \\ \| & & \| \\ {}^{II}E^2_{p,q} & & H''_pH'_{q,p}(M) \end{array}$$

Notational convention: Let M be a double (chain) complex of the third quadrant.
Set $M^{p,q} := M_{-p,-q}$, $p \geq 0$, $q \geq 0$, analogously for the differentials.

$E_r^{p,q} := E^r_{-p,-q}$, $p \geq 0$, $q \geq 0$, $r \geq 1$

$d_r : E_r \to E_r$ has now bidegree $(r,1-r)$. All filtrations now become decreasing: $F_{p+1} \subset F_p$! We thus may treat spectral sequences of double cochain complexes of the first quadrant as double chain complexes of the quadrant.

Proposition 1.4.13 Let $(E^r,d^r)_{r\geq 1}$ be one of the two spectral sequences of a double complex M of the first or third quadrant, and suppose $E^2_{p,q} = 0$ for $q \neq 0$ (the spectral sequence "degenerates").
Then $E^\infty_{p,q} = E^2_{p,q}$ for all $(p,q) \in \mathbb{Z}^2$ and $H_n(\mathrm{Tot}(M)) = E^2_{n,o}$, $n \in \mathbb{Z}$.

Proof. Consider $d^r_{p,q} : E^r_{p,q} \to E^r_{p-r,q+r-1}$
We have $d^r_{p,q} = 0$ for $r \geq 2$ and all $(p,q) \in \mathbb{Z}^2$ (since either $E^r_{p,q} = 0$ or $E^r_{p-r,q+r-1} = 0$), hence $E^r = \mathrm{Ker}\, d^r = \mathrm{Ker}\, d^r/\mathrm{Im}\, d^r = E^{r+1}$ for $r \geq 2$ and thus $E^\infty = E^2$.
Let M now be of the first quadrant, and let $T = \mathrm{Tot}(M)$ be the total complex of M. Then we have for either filtration:

$$F^{-1}T = 0$$

$$F^oT = F^oT/F^{-1}T = \begin{cases} (M_{o,*},d''_{o,*}) & \text{for the first filtration} \\ (M_{*,o},d'_{*,o}) & \text{for the second filtration} \end{cases}$$

$$(F^nT)_n = T_n.$$

Recall that $\phi^pH_n(T) = \mathrm{Im}(H_n(F^pT) \to H_n(T))$. Hence we obtain the finite chain

$$0 = \phi^{-1}H_n(T) \subset\cdots\subset \phi^nH_n(T) = H_n(T)$$

(since $H_n(F^nT) \to H_n(T)$ is surjective).

We know already that $E^2_{p,q} \simeq \phi^pH_n(T)/\phi^{p-1}H_n(T)$, $(n = p+q)$.
But only $E^2_{n,o}$ does not necessarily vanish. This implies $\phi^pH_n(T) = \phi^{p-1}H_n(T) = 0$ for $p < n$, and consequently $E^2_{n,o} = \phi^nH_n(T) = H_n(T)$.
This third-quadrant case is treated similarly.

I.2 Cyclic (co)homology and Hochschild (co)homology.

I.2.1 The double complex $C(A)$.

Let k be a unitary commutative ring, A an associative k-algebra (with unit), $A^e = A \otimes_k A^{op}$ the enveloping algebra of A, where A^{op} means the opposite algebra of A (with multiplication $a^o b^o = (ba)^o$). Note that an A-A-bimodule M (mixed associativity for the left and right actions, symmetric action of k) is equivalently a left or right A^e-module by the formulas

$$(a \otimes b^o)m = (am)b = a(mb) = m(b \otimes a^o).$$

In particular, A is naturally a left A^e-module, and the mapping $A^e \ni a \otimes b^o \mapsto ab \in A$ is an A^e-epimorphism.

Notation: $A^n := A \otimes_k A \otimes_k \cdots \otimes_k A$ (n times), $n \geq 1$.

$(a_1, \cdots, a_n) := a_1 \otimes \cdots \otimes a_n$

We shall consider every A^n, $n \geq 1$, as an A-A-bimodule in the following way: $a(a_1, \cdots, a_n)b = (aa_1, a_2, \cdots, a_{n-1}, a_n b)$ (left and right action on the external factors).

<u>Definition 2.1.1</u> The operators $b', b: A^{n+1} \to A^n$.

$$b'(a_o, \cdots, a_n) = \sum_{i=0}^{n-1} (-1)^i (a_o, \cdots, a_i a_{i+1}, \cdots, a_n)$$

$$b(a_o, \cdots, a_n) = \sum_{i=0}^{n-1} (-1)^i (a_o, \cdots, a_i a_{i+1}, \cdots, a_n) + (-1)^n (a_n a_o, \cdots, a_{n-1})$$

<u>Remark 2.1.2</u>

(1) The chain complex $\xrightarrow{b'} A^3 \xrightarrow{b'} A^2 \xrightarrow{b'} A$ is an acyclic complex of A^e-homomorphisms (the standard Hochschild resolution of A over A^e), since $s : A^n \to A^{n+1}$, defined by $s(a_1, \cdots, a_n) = (1, a_1, \cdots, a_n)$, is a homotopy operator (satisfying $b's + sb' = id$). (cf. [C.E., p.174]). When A is flat over k, we get an A^e-flat resolution of A.

Notation: (A^{*+1}, b'), the <u>acyclic Hochschild complex</u>.

(2) The chain complex $\xrightarrow{b} A^3 \xrightarrow{b} A^2 \xrightarrow{b} A$ is a complex of k-homomorphisms, which may be identified with $(A \otimes_{A^e} A^{*+2}, 1 \otimes b')$, augmented by $A^e \to A$.

More explicitely:

$$(A \otimes_{A^e} A^{*+2})_n = A \otimes_{A^e} A^{n+2} = A \otimes_{A^e} A^e \otimes_k A^n = A \otimes_k A^n = A^{n+1}$$

with the identification:

$$a \otimes_{A^e} (a_o, a_1, \cdots, a_n, a_{n+1}) = (a_{n+1} a a_o) \otimes_k (a_1, \cdots, a_n)$$

(which gives immediately $b = 1 \otimes b'$).

Notation: (A^{*+1}, b), the <u>Hochschild complex</u>.
$H_*(A) = H(A^{*+1}, b)$, the <u>Hochschild homology</u> of A.
Note that $H_n(A)$ is a subquotient of A^{n+1}. When A is k-flat (in particular, when k is a field), we get $H_n(A) = \mathrm{Tor}_n^{A^e}(A,A)$, $n \geq 0$.

<u>Remark 2.1.3</u> Homology of finite cyclic groups. Let G_n be "the" cyclic group of order n (think multiplicatively: n^{th} roots of unity), with generator $t = t_n$. There are two distinguished elements in $\mathbb{Z}[G_n]$, the group algebra of G_n over $\mathbb{Z}$:

$$D = 1 - t, \quad N = 1 + t + t^2 + \cdots + t^{n-1}$$

(1) $\mathbb{Z} \xleftarrow{\varepsilon} \mathbb{Z}[G_n] \xleftarrow{D} \mathbb{Z}[G_n] \xleftarrow{N} \mathbb{Z}[G_n] \xleftarrow{D}$

is a $\mathbb{Z}[G_n]$-free resolution of $\mathbb{Z}$
(where $\varepsilon(\sum_{i=0}^{n-1} z_i t^i) = \sum_{i=0}^{n-1} z_i$, and where D, N means multiplication by D, N). (cf. [C.E., p.251] or [Ro, p.296])

(2) For a left G_n-module M we set

$$H_m(G_n, M) = \mathrm{Tor}_m^{\mathbb{Z}[G_n]}(\mathbb{Z}, M), \quad m \geq 0.$$

$H_m(G_n, M)$ is thus the mth homology group of the chain complex

$$\begin{array}{ccccccc}
\mathbb{Z}[G_n] \otimes_{\mathbb{Z}[G_n]} M & \xleftarrow{D\otimes 1} & \mathbb{Z}[G_n] \otimes_{\mathbb{Z}[G_n]} M & \xleftarrow{N\otimes 1} & \mathbb{Z}[G_n] \otimes_{\mathbb{Z}[G_n]} M & \xleftarrow{D\otimes 1} & \\
\| & & \| & & \| & & \\
M & \xleftarrow{\quad D \quad} & M & \xleftarrow{\quad N \quad} & M & \xleftarrow{\quad D \quad} &
\end{array}$$

(3) For an associative k-algebra A we shall always consider A^n as a left G_n-module by letting the generator t act as the operator

$$t.(a_1, \cdots, a_n) = (-1)^{n-1}(a_n, a_1, \cdots, a_{n-1}).$$

Suppose now $\mathbb{Q} \subset k$, and let A^n_c be the G_n-submodul of A^n which consists of the cyclic (i.e. G_n-invariant) tensors. Then $A^n = A^n_c \oplus DA^n$ ($\pi = \frac{1}{n}N$ is the projection on A^n_c, and $\pi' = 1 - \frac{1}{n}N$ is the projection on $DA^n = (1-t)A^n$).

<u>Definition 2.1.4</u> The double complex $C(A)$.
Let $C(A) = (C_{p,q})_{p,q\geq 0}$ be the following double chain complex of the first quadrant (with differentials as indicated):

$$\begin{array}{lcccccc}
C(A)_{2,*}: & A^3 & \xleftarrow{\;D\;} & A^3 & \xleftarrow{\;N\;} & A^3 & \xleftarrow{\;D\;} \\
 & \downarrow b & & \downarrow -b' & & \downarrow b & \\
C(A)_{1,*}: & A^2 & \xleftarrow{\;D\;} & A^2 & \xleftarrow{\;N\;} & A^2 & \xleftarrow{\;D\;} \\
 & \downarrow b & & \downarrow -b' & & \downarrow b & \\
C(A)_{o,*}: & A & \xleftarrow{\;D\;} & A & \xleftarrow{\;N\;} & A & \xleftarrow{\;D\;} \\
 & & & & & & \\
 & C(A)_{*,o} & & C(A)_{*,1} & & C(A)_{*,2} &
\end{array}$$

Note that the even degree columns are Hochschild complexes whereas the odd degree columns are acyclic Hochschild complexes (up to a change of sign of the differential). The rows are the standard complexes for the homology of the various G_n (acting on A^n). We have actually a double chain complex by virtue of

<u>Lemma 2.1.5</u> $bD = Db'$ and $b'N = Nb$

<u>Proof</u>. Define $j\colon A^{n+1} \to A^n$ by $j(a_o,a_1,\cdot\cdot,a_n) = (-1)^n(a_na_o,a_1,\cdot\cdot,a_{n-1})$

Then it is easy to verify that

$$t^K_n \circ j \circ t^{-K-1}_{n+1}(a_o,\cdots,a_n) = (-1)^K(a_o,\cdots,a_Ka_{K+1},\cdots,a_n)$$

and hence $b' = \sum_{K=0}^{n=1} t^K \circ j \circ t^{-K-1}$, $b = \sum_{K=0}^{n} t^K \circ j \circ t^{-K-1}$.

This yields the desired equalities; for example

$b'N = NjN = Nb$.

<u>Remark 2.1.6</u> Denote, as usual, d' the vertical differential and d'' the horizontal differential of $C(A)$. Then we have (recall the notation of 1.4.10):

(a) $H'_{p,q}(C(A)) = \begin{cases} H_p(A) & \text{for } q \text{ even} \\ 0 & \text{for } q \text{ odd} \end{cases}$

(b) $H''_{p,q}(C(A)) = H_q(G_{p+1}, A^{p+1})$

Consider $(\mathrm{Tot}(C(A)),d)$, the total complex of $C(A)$. We have explicitely:

$$\begin{array}{cccccccccccc} \mathrm{Tot}(C(A))_n & = & A^{n+1} & \oplus & A^n & \oplus & A^{n-1} & \oplus\cdots\oplus & A^2 & \oplus & A \\ & & \| & & \| & & \| & & \| & & \| \\ & & C_{n,0} & \oplus & C_{n-1,1} & \oplus & C_{n-2,2} & \oplus\cdots\oplus & C_{1,n-1} & \oplus & C_{0,n} \end{array}$$

$$d_n = b_n \oplus (D_n - b'_{n-1}) \oplus (N_{n-1} + b_{n-2}) \oplus \cdots$$

I.2.2. The cyclic homology of an associative algebra.

Definition 2.2.1 Let A be a unitary associative k-algebra. $HC_*(A)$, the cyclic homology of A, is defined by

$HC_n(A) = H_n(\mathrm{Tot}(C(A)))$, $n \geq 0$.

There are two immediate consequences of the definition.

Proposition 2.2.2 (Extension of scalars)
Let A be a unitary associative algebra over k, and let $k \to K$ be a flat homomorphism of commutative rings. Then we have for ${}_KA = K \otimes_k A$:

$HC_n({}_KA) = K \otimes_k HC_n(A)$, $n \geq 0$.

(Attention: In general $HC_*(K \otimes_k A) \neq K \otimes_k HC_*(A)$ for the k-algebra $K \otimes_k A$!)

Proof. We shall use the well-known (cf. [Go, p.21])

Lemma: Let $k \to K$ be flat as above, (X,d) a differential k-module; then we have for the differential K-module $(K \otimes_k X, 1 \otimes d)$: $H(K \otimes_k X) = K \otimes_k H(X)$.

Now, $({}_KA)^n = (K \otimes_k A) \otimes_K (K \otimes_k A) \otimes_K \cdots \otimes_K (K \otimes_k A)$ (n times)

$= K \otimes_k A^n$ (associativity of the tensor product)

This yields $C({}_K A) = K \otimes_k C(A)$ (together with the differentials which are extended, too), and thus clearly

$$(\mathrm{Tot}(C({}_K A)), {}_K d) = (K \otimes_k \mathrm{Tot}(C(A)),\ 1 \otimes d).$$

Our lemma provides the desired result.

Note that the above proposition shows in particular that in the situation $\mathbb{Q} \subset k \subset \mathbb{C}$ vanishing or rank (dimension) questions about cyclic homology groups can be settled via complexification.

Remark 2.2.3 (Direct limits).
Let (k_i, α_{ij}) be a direct system of commutative rings, and let (A_i, φ_{ij}) be a direct system of unitary associative k_i-algebras such that the α_{ij} and the φ_{ij} are compatible. Set $A = \varinjlim A_i$, $k = \varinjlim k_i$. Suppose that A is a unitary associative k-algebra.
We have $A^n = \varinjlim A_i^n$ (cf. [Go, p.10]) and obtain easily

$$(C(A),\ d', d'') = (\varinjlim C(A_i),\ \varinjlim d_i',\ \varinjlim d_i'')$$

and finally $(\mathrm{Tot}C(A), d) = \varinjlim \mathrm{Tot}C(A_i),\ \varinjlim d_i)$.

Since $\varinjlim$ preserves exactness, the same argument which proves the lemma for 2.2.2 gives the

Lemma: Let (k_i, α_{ij}) and k be as above, and let $((X_i, d_i), f_{ij})$ be a direct system of differential k_i-modules (with the obvious compatibility conditions).
Then $(X, d) = (\varinjlim X_i,\ \varinjlim d_i)$ is a differential k-module and $H(X,d) = \varinjlim H(X_i, d_i)$.
So we may conclude that $HC_*(A) = \varinjlim HC_*(A_i)$.

Remark 2.2.4 The spectral sequences associated with the two standard filtrations of $\mathrm{Tot}C(A)$.
Recall 2.1.6; this reads in the language of the two spectral sequences of $\mathrm{Tot}(C(A))$:

$${}^{I}E^1_{p,q} = H''_{p,q}(C(A)) = H_q(G_{p+1}, A^{p+1}) \quad \text{(group homology)}$$

$${}^{II}E^1_{p,q} = H'_{q,p}(C(A)) = \begin{cases} H_q(A) & \text{for } p \text{ even} \\ 0 & \text{for } p \text{ odd} \end{cases}$$

(algebra homology)

We already know (1.4.9) that these two spectral sequences converge:

$$ {}^{I}E^2_{p,q} \underset{p}{\Rightarrow} H_n(\mathrm{Tot}(\mathcal{C}(A))) = HC_n(A), \quad n = p + q $$

$$ {}^{II}E^2_{p,q} \underset{p}{\Rightarrow} H_n(\mathrm{Tot}(\mathcal{C}(A))) = HC_n(A) $$

Remark 2.2.5

(1) Consider $\mathcal{C}(A)_{*,o} = (A^{*+1}, b)$, the first column of $\mathcal{C}(A)$ (which is simply the Hochschild complex for A). $DC(A)_{*,o}$ is a subcomplex of $\mathcal{C}(A)_{*,o}$ (since $bD = Db'$). Hence we have a quotient complex

$$ C^{\lambda}_{*}(A) = (\mathcal{C}(A)_{*,o}/D\mathcal{C}(A)_{*,o}, b) = (A^{*+1}/(1-t), b) $$

with $C^{\lambda}_{n}(A) = A^{n+1}/(1-t_{n+1})A^{n+1}$, $n \geq 0$.

When $\mathbb{Q} \subset k$ we already observed that A^{n+1}_{c}, the module of cyclic (G_{n+1}-invariant) tensors, parametrizes $C^{\lambda}_{n}(A)$.

(2) We have a natural epimorphism of chain complexes

$$ \rho : \mathrm{Tot}(\mathcal{C}(A)) \to C^{\lambda}_{*}(A) $$

$$ \rho_n : \mathrm{Tot}(\mathcal{C}(A))_n = A^{n+1} \oplus A^n \oplus \cdots \oplus A^2 \oplus A \downarrow A^{n+1}/(1-t) = C^{\lambda}_{n}(A) $$

which gives $HC_*(A) \xrightarrow{\rho} H^{\lambda}_{*}(A)$, where $H^{\lambda}_{*}(A)$ is the homology of $C^{\lambda}_{*}(A)$. Note that here (and in the sequel) we shall write the same symbol for a morphism of complexes and the morphism induced in homology.

Proposition 2.2.6 When $\mathbb{Q} \subset k$, then $\rho: \mathrm{Tot}(\mathcal{C}(A)) \to C^{\lambda}_{*}(A)$ is a quasi-isomorphism (i.e. induces an isomorphism in homology).

Proof.

$$ C^{\lambda}_{n}(A) = A^{n+1}/(1-t)A^{n+1} = H_o(G_{n+1}, A^{n+1}) = H''_{n,o}(\mathcal{C}(A)) $$

Thus $H^{\lambda}_{n}(A) = H'_n H''_{n,o}(\mathcal{C}(A)) = {}^{I}E^2_{n,o}$ (1.4.11)

Recall that for the first filtration of $\mathrm{Tot}(\mathcal{C}(A))$

$$ {}^{I}E^1_{p,q} = H''_{p,q}(\mathcal{C}(A)) = H_q(G_{p+1}, A^{p+1}) $$

Since $\mathbb{Q} \subset k$, we get ${}^{I}E^1_{p,q} = 0$ for $q > 0$ by the standard

Lemma: Let G be a finite group of order m. For every G-module M

and every $q > 0$: $mH_q(G,M) = 0$. (cf. [Ro, p.292]).

Thus a fortiori ${}^{I}E^2_{p,q} = 0$ for $q > 0$, and

$$HC_n(A) = H_n(\mathrm{Tot}(C(A))) \simeq {}^{I}E^2_{n,o} = H^{\lambda}_n(A),\ n \geq 0$$

by 1.4.13.

We have to convince ourselves that this isomorphism is actually induced by ρ. Write T for $\mathrm{Tot}(C(A))$ and suppress the upper "I". For r sufficiently large we had the isomorphism

$$D^r_{n+r-1,-r+1} \xrightarrow{\beta^r} E^r_{n,o} = E^2_{n,o}$$

where

$$D^r_{n+r-1,-r+1} = \mathrm{Im}(H_n(F^nT) \to H_n(F^{n+r-1}T)) = \phi^n H_n(T) = H_n(T).$$

By the definition of β^r

$$\beta^r[x_{n+1} \oplus \cdots \oplus x_1]_{H_n(T)} = [\beta_{n,o}[x_{n+1} \oplus \cdots \oplus x_1]_{H_n(F^nT)}].$$

But

$$\beta_{n,o}: D_{n,o} = H_n(F^nT) \to E_{n,o} = H_n(F^nT/F^{n-1}T) = A^{n+1}/(1-t)A^{n+1}$$

$$\beta_{n,o}[x_{n+1} \oplus \cdots \oplus x_1] = x_{n+1} \mathrm{mod}(1-t)$$

and thus

$$\beta^r[x_{n+1} \oplus \cdots \oplus x_1]_{H_n(T)} = [x_{n+1} \mathrm{mod}(1-t)] \in E^2_{n,o}$$

which gives the desired result.

I.2.3 Generalities about Mixed Complexes.

Definition 2.3.1 A mixed complex (M,b,B) is a non-negatively graded k-module $(M_n)_{n\geq o}$ together with a degree -1 endomorphism b and a degree $+1$ endomorphism B such that

$$b^2 = B^2 = [B,b] = 0$$

([,] stands for graded commutator: $[B,b] = Bb+bB$)

<u>Remark 2.3.2</u> (M,b) is a chain complex, (M,B) is a cochain complex. Morphisms of mixed complexes have to commute with both differentials.

<u>Definition 2.3.3</u> The associated chain complex $({}_BM,d)$ of a mixed complex (M,b,B) is defined by

$${}_BM_n = M_n \oplus M_{n-2} \oplus M_{n-4} \oplus \cdots$$

$$d_n(m_n,m_{n-2},m_{n-4},\cdots) = (bm_n+Bm_{n-2},bm_{n-2}+Bm_{n-4},\cdots)$$

(in short: $d = b+B$).

<u>Remark 2.3.4</u> For any chain complex $C = ((C_n)_{n\in\mathbb{Z}},d)$ and every $k \in \mathbb{Z}$ define the shifted chain complex

$$C[K] \quad \text{by} \quad C[K]_n = C_{n-K}$$
$$d[K]_n = d_{n-K}$$

We obtain the following exact sequence of chain complexes

$$0 \longrightarrow (M,b) \xrightarrow{I} ({}_BM,d) \xrightarrow{S} ({}_BM[2],d[2]) \longrightarrow 0$$

which reads in degree n simply

$$0 \to M_n \to M_n \oplus M_{n-2} \oplus M_{n-4} \oplus \cdots \to M_{n-2} \oplus M_{n-4} \oplus \cdots \to 0$$

S means projection (kill the first factor).

<u>Definition 2.3.5</u> Let (M,b,B) be a mixed complex.

$H_*(M) = H_*(M,b)$ the <u>homology</u> of (M,b,B)

$HC_*(M) = H_*({}_BM,d)$ the <u>cyclic homology</u> of (M,b,B)

<u>Proposition 2.3.6</u> There is a long exact homology sequence

$$\longrightarrow H_n(M) \xrightarrow{I} HC_n(M) \xrightarrow{S} HC_{n-2}(M) \xrightarrow{B} H_{n-1}(M) \longrightarrow$$

where the connecting homomorphism is induced by B.

<u>Proof</u>. Our exact sequence of chain complexes (2.3.4) gives rise to the above long exact homology sequence. We need only identify the connecting homomorphism.

In degree n and $n-1$ we have

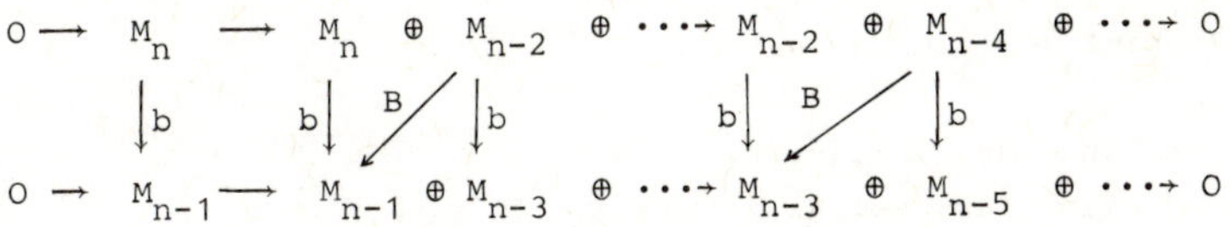

$$\begin{array}{ccccccccc}
0 \longrightarrow & M_n & \longrightarrow & M_n \oplus M_{n-2} \oplus \cdots & \rightarrow & M_{n-2} \oplus M_{n-4} \oplus \cdots & \rightarrow 0 \\
& \downarrow b & & b\downarrow \ \overset{B}{\swarrow} \ \downarrow b & & b\downarrow \ \overset{B}{\swarrow} \ \downarrow b & \\
0 \rightarrow & M_{n-1} & \longrightarrow & M_{n-1} \oplus M_{n-3} \oplus \cdots & \rightarrow & M_{n-3} \oplus M_{n-5} \oplus \cdots & \rightarrow 0
\end{array}$$

By definition of the connecting homomorphism ∂ we have $\partial[m_{n-2} \oplus \cdots]$ $= [Bm_{n-2}]$ (homology classes), which gives the desired result.

Complement 2.3.7 We have in lowest degrees

(i) an isomorphism $H_o(M) \xrightarrow{I} HC_o(M)$

(ii) an exact sequence $HC_o(M) \xrightarrow{B} H_1(M) \xrightarrow{I} HC_1(M) \longrightarrow 0$

(i.e. $HC_1(M)$ is a quotient of $H_1(M)$).

Remark-Definition 2.3.8 Recall the generalities on differential graded (d.g.) algebras and differential graded (d.g.) modules ([ML, p.189]). (Λ,d) is a d.g. k-algebra whenever (Λ,d) is a non-negatively graded chain complex which is a graded k-algebra such that the differential d is a graded k-derivation:

$$d(\lambda_1\lambda_2) = (d\lambda_1)\lambda_2 + (-1)^{|\lambda_1|}\lambda_1(d\lambda_2)$$

(where $|\cdot|$ means degree)

(M,b) is a d.g. Λ-module (Λ operating on the left), whenever (M,b) is a non-negatively graded chain complex over k which is a graded left Λ-module such that $b(\lambda m) = (d\lambda)m + (-1)^{|\lambda|}\lambda(bm)$ (this simply means that

$$0 \to (M,b) \to (\Lambda \oplus M, d \oplus b) \to (\Lambda,d) \to 0$$

is a trivial infinitesimal extension of d.g. k-algebras, where M, considered as a symmetric Λ-Λ-bimodule in the graded sense - $\lambda m = (-1)^{|\lambda||m|}m\lambda$ - becomes a graded ideal of square 0 in the middle term).

Consider now $\Lambda = k[\varepsilon]$, the ring of dual numbers over k, as a d.g. k-algebra with zero differential:

$$\Lambda_o = k, \quad \Lambda_1 = k\varepsilon, \quad \Lambda_i = 0 \quad \text{for} \quad i \geq 2; \quad d \equiv 0.$$

A d.g. Λ-module (M,b) is precisely a mixed complex (M,b,B), where B means left multiplication by ε:
The relevant relations $B^2 = [B,b] = 0$ simply translate to $\varepsilon^2 = 0$ and

$b(\varepsilon m) = (-1)^{|\varepsilon|}\varepsilon b(m) = -\varepsilon b(m)$.

<u>Remark 2.3.9</u> The functor $\mathrm{Tor}_*^\Lambda(-,-)$.
We shall only consider our special situation ($\Lambda = k[\varepsilon]$).

(i) Recall first the basic facts about tensor products of d.g. Λ-modules (cf. [ML, p.186,190]): Let (L,γ) be a d.g. right Λ-module, (M, b) a d.g. left Λ-module. Note the following particularity:

$b(\varepsilon m) = -\varepsilon b(m)$, but $\gamma(\ell\varepsilon) = \gamma(\ell)\varepsilon$

(think of the infinitesimal extension argument:

$0 \to (L,\gamma) \to (\Lambda \oplus L,\ 0 \oplus \gamma) \to (\Lambda,0) \to 0$

and of the symmetric Λ-Λ-bimodule structure of L in the graded sense) $L \underset{\Lambda}{\otimes} M$ is a d.g. k-module (chain complex) in the following way:
$(L \underset{\Lambda}{\otimes} M)_n$ is the k-module quotient of $(L \otimes M)_n = \sum\limits_{p+q=n} L_p \otimes M_q$ by the submodule generated by all differences $\ell\varepsilon \otimes m - \ell \otimes \varepsilon m$ in $(L \otimes M)_n = \sum\limits_{p+q=n} L_p \otimes M_q$.
The differential ∂ on $L \otimes M$ defined by

$\partial(\ell \otimes m) = \gamma(\ell) \otimes m + (-1)^{|\ell|}\ell \otimes bm$

passes to the graded quotient:

$$\partial(\ell\varepsilon \otimes m - \ell \otimes \varepsilon m) = (\gamma(\ell)\varepsilon \otimes m - \gamma(\ell) \otimes \varepsilon m) + (-1)^{|\ell|+1}(\ell\varepsilon \otimes bm - \ell \otimes \varepsilon b(m))$$

i.e. we may consider $\partial : (L \underset{\Lambda}{\otimes} M)_n \to (L \underset{\Lambda}{\otimes} M)_{n-1}$, as claimed.

$(- \underset{\Lambda}{\otimes} -)$ becames a covariant biadditive bifunctor on the categories of d.g. right Λ-modules and d.g. left Λ-modules (relative to morphisms of degree 0) to d.g. k-modules, which is right exact in both variables.

(ii) Now, let $\mathcal{L} : \cdots \to L_2 \overset{\delta}{\to} L_1 \overset{\delta}{\to} L_0$

be a d.g. Λ-projective resolution of L: Every L_p, $p \geq 0$, is a d.g. right Λ-module (with differential γ, say), which is projective in the category of d.g. right Λ-modules (cf. [ML, p.195]), and δ is a sequence of homomorphisms of d.g. right Λ-modules (chain transformations of degree 0), such that $L = L_0/\mathrm{Im}\,\delta$.
$\mathcal{L} \underset{\Lambda}{\otimes} M$ is a double complex in the following way: First, every $L_p \underset{\Lambda}{\otimes} M$

is a d.g. k-module (chain complex) as described above (i).
We obtain a double complex $(L \otimes_\Lambda M, \delta \otimes 1, \partial_-)$, where $\delta \otimes 1$ is given by the resolving differential δ of L, and where ∂_- stands for the sequence of differentials on $L_p \otimes_\Lambda M$, $p \geq 0$, up to the usual $(-1)^p$-sign in order to assure anticommutativity of the following square

$$\begin{array}{ccc} (L_p \otimes_\Lambda M)_{q-1} & \xleftarrow{(-1)^p \partial} & (L_p \otimes_\Lambda M)_q \\ \downarrow \delta\otimes 1 & & \downarrow \delta\otimes 1 \\ (L_{p-1} \otimes_\Lambda M)_{q-1} & \xleftarrow{(-1)^{p-1} \partial} & (L_{p-1} \otimes_\Lambda M)_q \end{array}$$

We can consider the total complex $T = \mathrm{Tot}(L \otimes_\Lambda M)$ associated with our double complex.
Explicitely:

$$T_n = \bigoplus_{p+q=n} (L_p \otimes_\Lambda M)_q$$

$d_n: T_n \to T_{n-1}$ is given by

$$d_n/(L_p \otimes_\Lambda M)_q = \delta \otimes 1 + (-1)^p \partial$$

(iii) Define $\mathrm{Tor}_*^\Lambda(L,M)$ by

$$\mathrm{Tor}_n^\Lambda(L,M) = H_n(T),\ n \geq 0.$$

Then we have by standard arguments:

(a) $\mathrm{Tor}_*^\Lambda(L,M)$ does not depend on the d.g. Λ-projective resolution chosen for L (you should rigorously work in the category of isomorphism classes of graded k-modules for homology).

(b) $\mathrm{Tor}_*^\Lambda(-,-)$ is a sequence of biadditive bifunctors which is exact connected in either variable (i.e. you have long exact homology sequences associated with short exact sequences of d.g. Λ-modules in either variable), and

$$\mathrm{Tor}_o^\Lambda(L,M) = L \otimes_\Lambda M.$$

(c) If you define symmetrically $\overline{\mathrm{Tor}}_*^\Lambda(L,M)$ using d.g. Λ-projective resolutions in the second variable, then there is a sequence of natural equivalences $\mathrm{Tor}_n^\Lambda(-,-) \simeq \overline{\mathrm{Tor}}_n^\Lambda(-,-)$ (dualize [H.St., p.144]; this avoids spectral sequence arguments).

$Tor_*^\Lambda(-,-)$ thus becomes a sequence of balanced bifunctors.

(iv) Caution: If you forget about the d.g. structures, considering L and M merely as right and left Λ-modules, then $Tor_*^\Lambda(L,M)$ also makes sense (as derived functor of the usual tensor product). You don't obtain the same homology, as is easily seen by the following example:

$L = k$ (considered as a trivial d.g. Λ-module via the augmentation map $\Lambda \to k$)

$$M = k \xleftarrow{0} k \xleftarrow{0} k \xleftarrow{0} k \xleftarrow{0} \cdots$$

(with ε acting as $+1$ shift)

You obtain in the d.g. context

$$Tor_n^\Lambda(L,M) = \begin{cases} k & n \text{ even} \\ 0 & n \text{ odd} \end{cases}$$

whereas in the standard context

$$Tor_n^\Lambda(L,M) = k \quad \text{for all} \quad n \geq 0.$$

Proposition 2.3.10 For every mixed complex (M,b,B) (equivalently: any d.g. Λ-module)

$$HC_*(M) = Tor_*^\Lambda(k,M)$$

(where k is the trivial Λ-module given by the augmentation $k[\varepsilon] \to k$).

Proof. Consider the following d.g. Λ-free resolution of k:

$$L: \quad \cdots \quad \Lambda[2] \xrightarrow{\varepsilon} \Lambda[1] \xrightarrow{\varepsilon} \Lambda$$

In degree p and $p-1$ we have thus

$$\begin{array}{lcccccccccccccc} \Lambda[p] & : & 0 & \leftarrow & 0 & \leftarrow & \cdots & 0 & \leftarrow & k & \leftarrow & k\varepsilon & \leftarrow & 0 & \leftarrow \cdots \\ \downarrow\varepsilon & & \downarrow & & \downarrow & & & \downarrow & & \downarrow\varepsilon & & \downarrow & & \downarrow & \\ \Lambda[p-1] & : & 0 & \leftarrow & 0 & \leftarrow & \cdots & k & \leftarrow & k\varepsilon & \leftarrow & 0 & \leftarrow & 0 & \leftarrow \cdots \\ & & & & & & & (p-1) & & (p) & & (p+1) & & & \end{array}$$

Let us look at the double complex $L \otimes_{\Lambda} M$ more closely:

$$(\Lambda[p] \otimes_{\Lambda} M)_q = (\Lambda[p] \otimes_{\Lambda} M)_q/\{\ell\varepsilon \otimes m - \ell \otimes \varepsilon m\}$$

$$= (1 \otimes M_{q-p} + \varepsilon \otimes M_{q-p-1})/\{\varepsilon \otimes m_{q-p-1} - 1 \otimes \varepsilon m_{q-p-1}\}$$

$$= M_{q-p}$$

$$\partial: (\Lambda[p] \otimes_{\Lambda} M)_q \to (\Lambda[p] \otimes_{\Lambda} M)_{q-1}$$

becomes after the above identification

$$\partial: M_{q-p} \longrightarrow M_{q-p-1}$$

$$m \mapsto (-1)^p b(m)$$

i.e. $\partial = (-1)^p b$ on $\Lambda[p] \otimes_{\Lambda} M = M[p]$.

Thus a typical square in our double complex

$(L \otimes_{\Lambda} M, \delta \otimes 1, \partial_-) = (L \otimes_{\Lambda} M, \varepsilon \otimes 1, \partial_-)$ looks like this:

$$\begin{array}{ccccc}
(\Lambda[p] \otimes_{\Lambda} M)_{q-1} = M_{q-p-1} & \xleftarrow{\partial_-=b} & (\Lambda[p] \otimes_{\Lambda} M)_q = M_{q-p} \\
\varepsilon \otimes 1 \downarrow \quad = B & & \varepsilon \otimes 1 \downarrow \quad = B \\
(\Lambda[p-1] \otimes_{\Lambda} M)_{q-1} = M_{q-p} & \xleftarrow{\partial_-=b} & (\Lambda[p-1] \otimes_{\Lambda} M)_q = M_{q-p+1}
\end{array}$$

For the associated total complex T we obtain

$$T_n = \bigoplus_{p+q=n} (\Lambda[p] \otimes M)_q = \bigoplus_{p+q=n} M_{q-p} = M_n \oplus M_{n-2} \oplus M_{n-4} \oplus \cdots$$

$$= ({}_BM)_n$$

$$d = b + B$$

Consequently $(T,d) = ({}_BM,d)$, which proves our assertion.

Remark 2.3.11 It is clear that a morphism of d.g. Λ-modules $F: (M,b) \to (N,b)$, i.e. a chain transformation which commutes with the B-operators, gives rise to a commutative diagram of complex homomorphisms

$$\begin{array}{ccccccccc}
0 & \longrightarrow & (M,b) & \longrightarrow & ({}_BM,d) & \longrightarrow & ({}_BM[2],d[2]) & \longrightarrow & 0 \\
 & & \Big\downarrow F & & \Big\downarrow {}_BF & & \Big\downarrow {}_BF[2] & & \\
0 & \longrightarrow & (N,b) & \longrightarrow & ({}_BN,d) & \longrightarrow & ({}_BN[2],d[2]) & \longrightarrow & 0
\end{array}$$

and thus to a commutative diagram relating the long exact sequences

$$\begin{array}{ccccccccc}
\longrightarrow & H_n(M) & \xrightarrow{I} & HC_n(M) & \xrightarrow{S} & HC_{n-2}(M) & \xrightarrow{B} & H_{n-1}(M) & \longrightarrow \\
 & \Big\downarrow H_n(F) & & \Big\downarrow HC_n(F) & & \Big\downarrow & & \Big\downarrow & \\
\longrightarrow & H_n(N) & \xrightarrow{I} & HC_n(N) & \xrightarrow{S} & HC_{n-2}(N) & \xrightarrow{B} & H_{n-1}(N) & \longrightarrow
\end{array}$$

We want to discuss under which milder assumptions of F we still obtain such a diagram.

Definition 2.3.12 Let (M,b,B) and (N,b,B) be two mixed complexes (d.g. Λ-modules).
A strongly homotopy Λ-map from M to N is a sequence $(G^{(i)})_{i\geq 0}$ of graded maps $G^{(i)}: M \rightarrow N$ of degree $2i$ for all $i \geq 0$ such that

(1) $G^{(o)}b = bG^{(o)}$, i.e. $G^{(o)}$ is a morphism from (M,b) to (N,b)

(2) $G^{(i)}B + G^{(i+1)}b = BG^{(i)} + bG^{(i+1)}$ for all $i \geq 0$.

Visualization of the connection between $G^{(o)}$ and $G^{(1)}$:

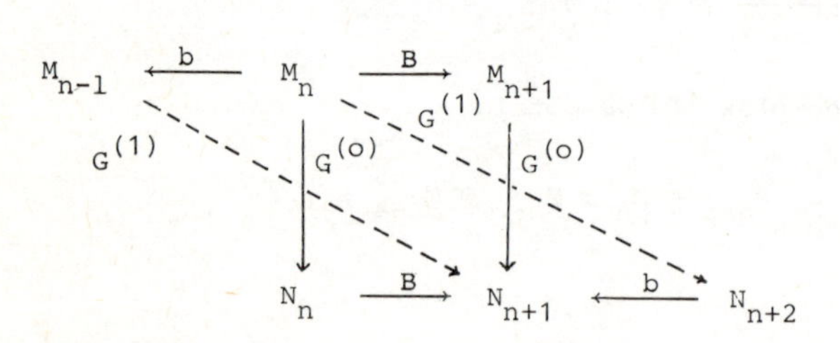

As maps from M_n to N_{n+1} the two following maps are equal:

$G^{(o)}B + G^{(1)}b = BG^{(o)} + bG^{(1)}$

Proposition 2.3.13 Let (M,b,B) and (N,b,B) be two mixed complexes; assume that there exists a strongly homotopy Λ-map $(G^{(i)})_{i\geq 0}$ from M to N. Then there exists a map of complexes $G: {}_BM \rightarrow {}_BN$ such that following diagram is commutative

$$\begin{array}{ccccccccc}
0 & \longrightarrow & (M,b) & \longrightarrow & ({}_BM,d) & \longrightarrow & ({}_BM[2],d[2]) & \longrightarrow & 0 \\
 & & \downarrow G^{(o)} & & \downarrow G & & \downarrow G[2] & & \\
0 & \longrightarrow & (N,b) & \longrightarrow & ({}_BN,d) & \longrightarrow & ({}_BN[2],d[2]) & \longrightarrow & 0
\end{array}$$

Proof. Look at the situation in degree n:

$$\begin{array}{ccccccccc}
({}_BM)_n & = & M_n & \oplus & M_{n-2} & \oplus & M_{n-4} & \oplus & \cdots \\
\downarrow G_n & & \downarrow G^{(o)} & \swarrow G^{(1)} & & \swarrow G^{(2)} & & & \\
({}_BN)_n & = & N_n & \oplus & N_{n-2} & \oplus & N_{n-4} & \oplus & \cdots
\end{array}$$

Define $G_n|M_{n-2i} = G^{(o)} + G^{(1)} + \ldots + G^{(i)}, \quad 0 \le i \le [\frac{n}{2}]$.

The compatibility of $G^{(o)}$ with the operators b, and the relations

$$G^{(o)}B + G^{(1)}b = BG^{(o)} + bG^{(1)}$$

$$G^{(1)}B + G^{(2)}b = BG^{(1)} + bG^{(2)}$$

$$\vdots$$

$$G^{(i)}B+G^{(i+1)}b = BG^{(i)} + bG^{(i+1)}$$

yield immediately the commutation of G with the total differentials $d = B+b$ (on ${}_BM$ and on ${}_BN$).
Hence G is a complex homomorphism, and trivially compatible with $G^{(o)}$ and $G[2]$ as asserted.

Consequence 2.3.14 In the situation of 2.3.13 we have a commutative diagram

$$\begin{array}{ccccccccc}
\longrightarrow H_n(M) & \xrightarrow{I} & HC_n(M) & \xrightarrow{S} & HC_{n-2}(M) & \xrightarrow{B} & H_{n-1}(M) & \longrightarrow \\
\downarrow H_n(G^{(o)}) & & \downarrow HC_n(G) & & \downarrow & & \downarrow & \\
\longrightarrow H_n(N) & \xrightarrow{I} & HC_n(N) & \xrightarrow{S} & HC_{n-2}(N) & \xrightarrow{B} & H_{n-1}(N) & \longrightarrow
\end{array}$$

Proposition 2.3.15 In the situation of 2.3.13 the following holds:

$G^{(o)}: M \to N$ is a quasi-isomorphism if and only if

$G : {}_BM \to {}_BN$ is a quasi-isomorphism

Proof. Recall the five-lemma (cf. [Bou, AX.7]):

Lemma: Given a diagram with exact rows

$$\begin{array}{ccccccccc} M_1 & \longrightarrow & M_2 & \longrightarrow & M_3 & \longrightarrow & M_4 & \longrightarrow & M_5 \\ \downarrow f_1 & & \downarrow f_2 & & \downarrow f_3 & & \downarrow f_4 & & \downarrow f_5 \\ N_1 & \longrightarrow & N_2 & \longrightarrow & N_3 & \longrightarrow & N_4 & \longrightarrow & N_5 \end{array}$$

we have

(i) f_2, f_4 injective, f_1 surjective $\Rightarrow$ f_3 injective

(ii) f_2, f_4 surjective, f_5 injective $\Rightarrow$ f_3 surjective

In particular: f_1, f_2, f_4, f_5 isomorphisms $\Rightarrow$ f_3 isomorphism

The assertion of the proposition follows by the five-lemma. The implication "$G^{(o)}$ quasi-isomorphism $\Rightarrow$ G quasi-isomorphism" is seen by induction on n:

$$\begin{array}{ccccccccc} HC_{n-1}(M) & \longrightarrow & H_n(M) & \longrightarrow & HC_n(M) & \longrightarrow & HC_{n-2}(M) & \longrightarrow & H_{n-1}(M) \\ \downarrow & & \downarrow & & \downarrow & & \downarrow & & \downarrow \\ HC_{n-1}(N) & \longrightarrow & H_n(N) & \longrightarrow & HC_n(N) & \longrightarrow & HC_{n-2}(N) & \longrightarrow & H_{n-1}(N) \end{array}$$

whereas the implication "G quasi-isomorphism $\Rightarrow$ $G^{(o)}$ quasi-isomorphism" follows directly from the five-lemma (every H_n in the long exact sequences has two HC-partners on the left and on the right).

I.2.4. Cyclic Homology and Hochschild Homology.

Remark-Definition 2.4.1 Recall the definition of the double complex $C(A)$ associated with a unital associative k-algebra A: The columns $C(A)_{*,2\nu}$ are the Hochschild complexes (A^{*+1}, b) the columns $C(A)_{*,2\nu+1}$ are the acyclic Hochschild complexes $(A^{*+1}, -b')$, $\nu \geq 0$. The homotopy operator $s: A^n \to A^{n+1}$ of the acyclic Hochschild complex $(s(a_1,\cdots,a_n)$

$= (1,a_1,\cdots,a_n))$ allows to define an operator B of degree +1 on the Hochschild complex (A^{*+1},b):

$$B = DsN : A^{n+2} \xleftarrow{D} A^{n+2}, \quad A^{n+2} \xleftarrow{s} A^{n+1}, \quad A^{n+1} \xleftarrow{N\cdot} A^{n+1}$$

Most explicitely:

$$\begin{aligned} B(a_o,a_1,\cdots,a_n) &= (1-t)s\sum_{i=0}^{n}(-1)^{in}(a_i,a_{i+1},\cdots,a_n,a_o,\cdots,a_{i-1}) \\ &= (1-t)\sum_{i=0}^{n}(-1)^{in}(1,a_i,\cdots,a_n,a_o,\cdots,a_{i-1}) \\ &= \sum_{i=0}^{n}(-1)^{in}(1,a_i,\cdots,a_n,a_o,\cdots,a_{i-1}) + \\ &\quad + \sum_{i=0}^{n}(-1)^{in}(a_i,1,a_{i+1},\cdots,a_{i-1}) \end{aligned}$$

<u>Proposition 2.4.2</u> $C(A) = (A^{*+1},b,B)$ is a <u>mixed complex</u>.

<u>Proof</u>. We have to establish the relations

$$B^2 = [B,b] = 0$$

$$B^2 = (1-t)sN(1-t)sN = 0, \text{ since } N(1-t) = 0$$

$$bB + Bb = b(1-t)sN + (1-t)sNb = (1-t)(b's+sb')N = (1-t)N = 0.$$

(recall 2.1.5)

<u>Remark 2.4.3</u> The homology $H_*(C(A))$ is the Hochschild homology $H_*(A)$ of A. The cyclic homology $HC_*(C(A))$ is the cyclic homology $HC_*(A)$ of A, as will be shown below (2.4.5).

<u>Remark 2.4.4</u> We have to look more closely at $({}_BC(A),d)$, the chain complex associated with our mixed Hochschild complex.

Consider ${}_BC(A)_n = A^{n+1} \oplus A^{n-1} \oplus A^{n-3} \oplus \ldots$ as a submodule of

$\mathrm{Tot}(C(A))_n = A^{n+1} \oplus A^n \oplus A^{n-1} \oplus \ldots$, hence ${}_BC(A)$ as a graded submodul

of $\mathrm{Tot}(C(A))$.

Define $f: {}_BC(A) \to Tot(C(A))$ by $f = id + sN$.

More precisely:

For $x \in A^{n+1-2i} \subset {}_BC(A)_n$ we have

$$f(x) = \begin{cases} x & \text{if } i = 0 \\ (sNx, x) \in A^{n+2-2i} \oplus A^{n+1-2i} & \text{if } i > 0 \end{cases}$$

$f({}_BC(A)_n) \subset Tot(C(A))_n$, and we have commutativity of the following diagram (i.e. f is a homomorphism of chain complexes):

$$\begin{array}{ccc} {}_BC(A)_n & \xrightarrow{f} & Tot(C(A))_n \\ \downarrow d=b+B & & \downarrow d=(b,D-b',b+N,\dots) \\ {}_BC(A)_{n-1} & \xrightarrow{f} & Tot(C(A))_{n-1} \end{array}$$

Let us verify commutativity.

Take $x \in A^{n+1-2i} \subset {}_BC(A)_n$.

For $i = 0$, $fd(x) = fb(x) = b(x) = bf(x) = df(x)$,

and for $i > 0$

$$df(x) = (D-b', b+N)(sNx, x) = (Bx, Nx-b'sNx, bx) = (Bx, sb'Nx, bx)$$

$$fd(x) = f(Bx, 0, bx) = (sNBx, Bx, sNbx, bx) = (Bx, sb'Nx, bx)$$

(since $NB = 0$, $Nb = b'N$ (2.1.5)).

<u>Proposition 2.4.5</u> $f: {}_BC(A) \to Tot(C(A))$ is a quasi-isomorphism, i.e. $H_n(f): HC_n(C(A)) \to HC_n(A)$ is an isomorphism for all $n \geq 0$.

<u>Proof</u>. Write $C(A) = (C_{s,t})_{s,t\geq 0}$ (where $C_{s,t} = A^{s+1}$, $s,t \geq 0$).

We have for $T = Tot(C(A))$ and for ${}_BC(A)$:

$$T_n = \bigoplus_{s+t=n} C_{s,t}, \quad {}_BC(A)_n = \bigoplus_{\substack{s+t=n \\ t \text{ even}}} C_{s,t}$$

In this notation $B: C_{s,t} \to C_{s+1,t-2}$.

Let us consider the second filtration of $\mathcal{T} = \mathrm{Tot}(\mathcal{C}(A))$ and the induced filtration on ${}_BC(A)$.

Explicitely:

$$(F^p{}_BC(A))_n = \bigoplus_{\substack{j\le p \\ j \text{ even}}} C_{n-j,j}, \qquad (F^p\mathcal{T})_n = \bigoplus_{j\le p} C_{n-j,j}$$

$F^p{}_BC(A)$ is a subcomplex of ${}_BC(A)$ for all $p \ge 0$ (since B lowers the second index).

$$f: {}_BC(A) \to \mathcal{T} = \mathrm{Tot}(\mathcal{C}(A))$$

becomes a homomorphism of filtered complexes.

(since $f(C_{q,p}) \subset C_{q+1,p-1} \oplus C_{q,p}$)

We have to look at the spectral sequences associated with these filtrations on ${}_BC(A)$ and $\mathcal{T} = \mathrm{Tot}(\mathcal{C}(A))$.

We know already:

$$E^1_{p,q}(\mathcal{T}) = H'_{q,p}(\mathcal{C}(A)) = \begin{cases} H_q(A) & \text{for } p \text{ even} \\ 0 & \text{for } p \text{ odd} \end{cases}$$

(see 2.2.4)

and that $E^2_{p,q}(\mathcal{T}) \underset{p}{\Rightarrow} H_n(\mathcal{T}) = HC_n(A)$, $n = p + q$.

On the other hand, the filtration on ${}_BC(A)$ is clearly bounded, and thus

$$E^2_{p,q}({}_BC(A)) \underset{p}{\Rightarrow} H_n({}_BC(A)) = HC_n(C(A)), \; n = p + q.$$

Let us calculate $E^1_{p,q}({}_BC(A))$.

$$E^1_{p,q}({}_BC(A)) = H_{p+q}(F^p{}_BC(A)/F^{p-1}{}_BC(A))$$

But $$(F^p{}_BC(A)/F^{p-1}{}_BC(A))_{p+q} = \begin{cases} C_{q,p} & p \text{ even} \\ 0 & p \text{ odd} \end{cases}$$

Since $B(F^p{}_B C(A)) \subset F^{p-1}{}_B C(A)$, the induced differential on $F^p{}_B C(A)/F^{p-1}{}_B C(A)$ is merely given by b. Thus

$$E^1_{p,q}({}_B C(A)) = \begin{cases} H_q(A) & p \text{ even} \\ 0 & p \text{ odd} \end{cases}$$

Consider now $f^1: E^1({}_B C(A)) \to E^1(T)$. f^1 is induced by f in homology, and f maps $x \in C_{q,p}$, p even, to $(sNx,x) \in C_{q+1,p-1} \oplus C_{q,p}$. Hence

$$f^1_{p,q}: \begin{cases} H_q(A) \to H_q(A) & p \text{ even} \\ 0 \to 0 & p \text{ odd} \end{cases}$$

is the identity.

By the approximation theorem 1.3.7 we can conclude that

$H(f): HC_*(C(A)) \to HC_*(A)$

is an isomorphism, which proves our assertion.

Theorem 2.4.6 For every unital associative k-algebra A there is a long exact sequence

$$\dots \to H_n(A) \xrightarrow{I} HC_n(A) \xrightarrow{S} HC_{n-2}(A) \xrightarrow{B} H_{n-1}(A) \to \dots$$

Proof. This is an immediate consequence of 2.3.6 and 2.4.5: the exact sequence of chain complexes

$$0 \longrightarrow (C(A),b) \longrightarrow ({}_B C(A),d) \longrightarrow ({}_B C(A)[2],d[2]) \longrightarrow 0$$

yields our long exact sequence when passing to homology and identifying $H_*(A) = H_*(C(A))$, $HC_*(A) = HC_*(C(A))$.

The connecting homomorphism is induced by our operator B.

Complement 2.4.7 Recall 2.3.7: We have in lowest degrees

(i) an isomorphism $0 \to H_0(A) \xrightarrow{I} HC_0(A) \to 0$

(ii) an epimorphism $H_1(A) \to HC_1(A) \to 0$

<u>Application 2.4.8</u> Let $M_r(k)$ be the k-algebra of r×r-matrices with coefficients in the commutative ring k.

$$\text{Then } HC_n(M_r(k)) = \begin{cases} k & \text{for } n \text{ even} \\ 0 & \text{for } n \text{ odd} \end{cases}$$

We have in Hochschild homology:

$$H_o(M_r(k)) = M_r(k)/[M_r(k),M_r(k)] = k$$

$$H_n(M_r(k)) = 0 \quad \text{for} \quad n \geq 1$$

($M_r(k)$ is k-free, hence $H_n(M_r(k)) = \mathrm{Tor}_n^{M_r(k)^e}(M_r(k),M_r(k))$ but $M_r(k)$ is $M_r(k)^e$-projective: [C.E., p.179])

Thus (2.4.7): $HC_o(M_r(k)) = k$

$HC_1(M_r(k)) = 0$

The long exact sequence (2.4.6) yields immediately $HC_n(M_r(k)) \simeq HC_{n-2}(M_r(k))$ for $n \geq 2$, and thus our result.

<u>Remark 2.4.9</u> Let A_1 and A_2 be two unital associative k-algebras, $G^{(o)}$: $(C(A_1),b) \to (C(A_2),b)$ a homomorphism of the Hochschild complexes such that there exists a homomorphism G: $({}_BC(A_1),d) \to ({}_BC(A_2),d)$ making the following diagram commutative:

$$\begin{array}{ccccccccc} 0 & \longrightarrow & (C(A_1),b) & \longrightarrow & ({}_BC(A_1),d) & \longrightarrow & ({}_BC(A_1)[2],d[2]) & \longrightarrow & 0 \\ & & \downarrow G^{(o)} & & \downarrow G & & \downarrow G[2] & & \\ 0 & \longrightarrow & (C(A_2),b) & \longrightarrow & ({}_BC(A_2),d) & \longrightarrow & ({}_BC(A_2)[2],d[2]) & \longrightarrow & 0 \end{array}$$

Then (2.3.15) $G^{(o)}$ is a quasi-isomorphism if and only if G is a quasi-isomorphism

($H_*(A_1) \simeq H_*(A_2)$ via $H_*(G^{(o)}) \Leftrightarrow HC_*(A_1) \simeq HC_*(A_2)$ via $HC_*(G)$)

<u>Caution</u>: The equivalence is only true as a statement <u>in all degrees</u>.

Note that every homomorphism of k-algebras $g: A_1 \to A_2$ gives rise to such a couple $(G^{(o)}, G)$ in an obvious way (functoriality of $H_*(-)$ and $HC_*(-)$).

Proposition 2.4.10 Let A_1, A_2 be two unital associative k-algebras, which are k-flat. Consider $A = A_1 \times A_2$, their direct product, and the projections

$\pi_1: A \to A_1$, $\pi_2: A \to A_2$.

Then $HC_*(A) \xrightarrow[HC(\pi_2)]{HC(\pi_1)} HC_*(A_1) \oplus HC_*(A_2)$

is an isomorphism.

Proof. It is well-known ([C.E., p.173]) that the result holds in Hochschild homology (the flatness assumption allows to identify Hochschild homology with a Tor-functor, cf. 2.1.2).
Consider now the following commutative diagram (we have dropped the arguments; the meaning is obvious):

$$\begin{array}{ccccccccccc} \longrightarrow & HC_{n-1} & \longrightarrow & H_n & \longrightarrow & HC_n & \longrightarrow & HC_{n-2} & \longrightarrow & H_{n-1} & \longrightarrow \\ & \downarrow\downarrow & & \downarrow\downarrow & & \downarrow\downarrow & & \downarrow\downarrow & & \downarrow\downarrow & \\ \rightrightarrows & HC^{(1)}_{n-1} \oplus HC^{(2)}_{n-1} & \rightrightarrows & H^{(1)}_n \oplus H^{(2)}_n & \rightrightarrows & HC^{(1)}_n \oplus HC^{(2)}_n & \rightrightarrows & HC^{(1)}_{n-2} \oplus HC^{(2)}_{n-2} & \rightrightarrows & H^{(1)}_{n-1} \oplus H^{(2)}_{n-1} & \rightrightarrows \end{array}$$

Once more, the five-lemma yields our result by induction on n.

Corollary 2.4.11 Let A be a unital associative flat k-algebra, $\tilde{A} = k1 \oplus A$ the unital k-algebra obtained by adjunction of a (new) unity. Then

$$HC_n(\tilde{A}) = HC_n(k) \oplus HC_n(A) = \begin{cases} k \oplus HC_n(A) & n \text{ even} \\ HC_n(A) & n \text{ odd} \end{cases}$$

Proof. Let $e \in A$ be the unity of A. Then

$\phi: k \times A \to \tilde{A}$

$(\alpha, a) \to (\alpha, a - \alpha e)$

is an isomorphism of k-algebras. 2.4.8 and 2.4.10 now give our result.

Application 2.4.12 Cyclic homology of Clifford algebras. For the moment , we shall only treat the nondegenerate case. Let k be a field of characteristic $\neq 2$, and let K be an algebraic closure of k. Let (V,Q) be a nondegenerate finite-dimensional quadratic space over k, $A = C(V,Q)$ the associated Clifford algebra. By extension of scalars one obtains

$$_KA = {_KC(V,Q)} = C({_KV}, {_KQ})$$

But, since K is algebraically closed,

$$_KA \simeq \begin{cases} M_{2^m}(K) & \dim V = 2m \\ M_{2^m}(K) \times M_{2^m}(K) & \dim V = 2m+1 \end{cases}$$

Thus

$$HC_n({_KA}) = \begin{cases} \left.\begin{cases} K & \dim V \text{ even} \\ K \oplus K & \dim V \text{ odd} \end{cases}\right\} & n \text{ even} \\ 0 & n \text{ odd} \end{cases}$$

But by 2.2.2: $HC_n({_KA}) = K \otimes HC_n(A)$, $n \geq 0$
Hence we obtain finally:

$$HC_n(C(V,Q)) = \begin{cases} \left.\begin{cases} k & \dim V \text{ even} \\ k \oplus k & \dim V \text{ odd} \end{cases}\right\} & n \text{ even} \\ 0 & n \text{ odd} \end{cases}$$

Definition 2.4.13 The double complex $\mathcal{B}(A)$.
Let A be a unital associative k-algebra, $\mathcal{C}(A) = (C_{p,q})_{p,q\geq 0}$ the double (Hochschild) complex associated with A.
Define a now double complex $\mathcal{B}(A)$ be deleting the acyclic columns in $\mathcal{C}(A)$ and rendering B a horizontal differential:

$$\mathcal{B}(A)_{p,q} = \begin{cases} C_{p-q,2q} & p \geq q \geq 0 \\ 0 & \text{otherwise} \end{cases}$$

Vertical differential: b

Horizontal differential: B

(Recall: $B: C_{p-q,2q} = A^{p-q+1} \to A^{p-q+2} = C_{p-q+1,2q-2}$, cf. 2.4.1)

$$\begin{array}{lccccccc}
\mathcal{B}(A)_{3,*}: & A^4 & \xleftarrow{B} & A^3 & \xleftarrow{B} & A^2 & \xleftarrow{B} & A \\
 & \downarrow b & & \downarrow b & & \downarrow b & & \downarrow \\
\mathcal{B}(A)_{2,*}: & A^3 & \xleftarrow{B} & A^2 & \xleftarrow{B} & A & \longleftarrow & 0 \\
 & \downarrow & & \downarrow & & \downarrow & & \downarrow \\
\mathcal{B}(A)_{1,*}: & A^2 & \xleftarrow{B} & A & \longleftarrow & 0 & \longleftarrow & 0 \\
 & \downarrow & & \downarrow & & \downarrow & & \downarrow \\
\mathcal{B}(A)_{0,*}: & A & \longleftarrow & 0 & \longleftarrow & 0 & \longleftarrow & 0 \\
 & & & & & & & \\
 & \mathcal{B}(A)_{*,0} & & \mathcal{B}(A)_{*,1} & & \mathcal{B}(A)_{*,2} & & \mathcal{B}(A)_{*,3}
\end{array}$$

<u>Remark 2.4.14</u> $(\mathrm{Tot}(\mathcal{B}(A)),d) = ({}_{B}C(A),d)$

i.e. $HC_n(A) = H_n(\mathrm{Tot}(\mathcal{B}(A)),d), \quad n \geq 0.$

<u>Theorem 2.4.15</u> Let $(E^r,d^r)_{r\geq 1}$ be the spectral sequence associated with the second filtration of $T = \mathrm{Tot}(\mathcal{B}(A))$.

Then $E^2_{p,q} \underset{p}{\Rightarrow} HC_n(A) \quad (n = p+q)$

and the following holds:

(1) $E^1_{p,q} = H_{q-p}(A), \quad q \geq p$

(2) $d^1_{p,q}: H_{q-p}(A) \to H_{q-p+1}(A)$

is induced by B

<u>Proof</u>. Writing $\mathcal{B}$ for $\mathcal{B}(A)$ we have

$$(F^p\mathcal{B}/F^{p-1}\mathcal{B})_{p+q} = \mathcal{B}_{q,p} = \begin{cases} A^{q-p+1} & q \geq p \\ 0 & \text{otherwise} \end{cases}$$

Thus $F^pB/F^{p-1}B = (B_{*,p},b) = (A^{*-p+1},b)$.

We obtain $E^1_{p,q} = H_{q-p}(A),\ q \geq p$.

It remains to identify $d^1_{p,q}: H_{q-p}(A) \to H_{q-p+1}(A)$.

$d^1_{p,q}$ is anyway the connecting homomorphism in the long exact homology sequence of

$$0 \longrightarrow F^{p-1}B/F^{p-2}B \longrightarrow F^pB/F^{p-2}B \longrightarrow F^pB/F^{p-1}B \longrightarrow 0$$

(see 1.2.2)

But in 1.4.11 we saw that for either standard spectral sequence of a double complex, this connecting homomorphism is induced by the differential still alive after having taken first step homology.
Let us write down once more the situation in degrees p+q and p+q-1:

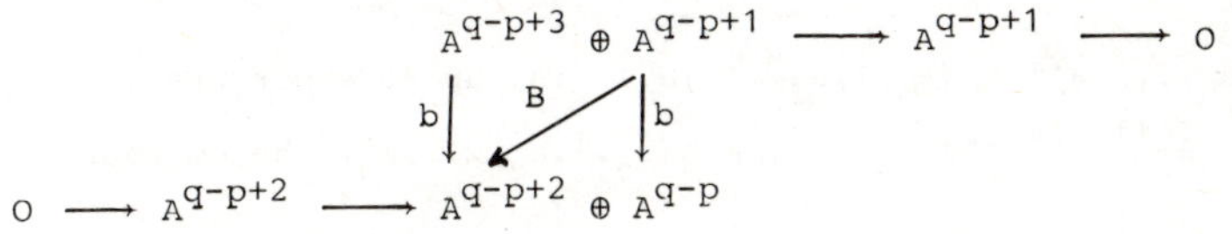

$$\begin{array}{ccccc} & & A^{q-p+3} \oplus A^{q-p+1} & \longrightarrow & A^{q-p+1} \longrightarrow 0 \\ & & b\downarrow \quad \swarrow B \quad \downarrow b & & \\ 0 \longrightarrow A^{q-p+2} & \longrightarrow & A^{q-p+2} \oplus A^{q-p} & & \end{array}$$

Note that $(F^pB/F^{p-1}B)_{p+q} = A^{q-p+1}$

$(F^{p-1}B/F^{p-2}B)_{p+q-1} = A^{q-p+2}$.

This concludes the proof of our theorem.

I.2.5 Nonunital and Reduced Cyclic Homology.

Remark-Definition 2.5.1 Let A be a unital associative k-algebra; consider the following subcomplex (D_*,b) of the Hochschild complex (A^{*+1},b):
$D_n \subset A^{n+1}$ is spanned by all elements $(a_o,a_1,\ldots,a_n)$ such that $a_i = 1$ for some $i: 1 \leq i \leq n$.
(note that $bD_n \subset D_{n-1}$, since for a typical element $(a_o,a_1, \quad ,a_n)$ of D_n the terms in $b(a_o,a_1,\ldots,a_n)$ where the argument 1 does not occur cancel out).

(1) (D_*,b) is acyclic.
The verification can be done by hand.

First, in lowest degree we have

$$(a_o,1) = b(a_o,1,1)$$

Then, consider an element of the form $(a_o,1,a_2,\dots,a_n) \in D_n$ such that

$$b(a_o,1,a_2,\dots,a_n) = (a_o,1,a_2\,a_3,\dots,a_n) - \dots + (-1)^n(a_na_o,1,\dots,a_{n-1}) = 0$$

(the first two terms have cancelled out).

I claim that

$$(a_o,1,a_2,\dots,a_n) = b(a_o,1,1,a_2,\dots,a_n)$$

Now, $b(a_o,1,1,a_2,\dots,a_n) = (a_o,1,a_2,\dots,a_n) - (a_o,1,1,a_2a_3,\dots,a_n)$

$$+ \dots + (-1)^{n+1}(a_na_o,1,1,a_2,\dots,a_{n-1}).$$

But $b(a_o,1,1,a_n,\dots,a_n) - (a_o,1,a_2,\dots,a_n)$ is, up to a permutation isomorphism of A^{n+1}, of the form $1 \otimes b(a_o,1,a_2,\dots,a_n)$, hence equal to zero.

The other cases ($a_i = 1$ for some $i \geq 2$) are treated similarly.

(2) With $\overline{A} = \mathrm{CoKer}(k \to A)$ we have

$$A^{n+1}/D_n = A \otimes \underbrace{\overline{A} \otimes \dots \otimes \overline{A}}_{n \text{ times}} = A \otimes \overline{A}^n$$

(right exactness of the tensor product).

We obtain a short exact sequence of chain complexes

$$0 \to (D_*,b) \to (A^{*+1},b) \to (A \otimes \overline{A}^*,b) \to 0$$

where $(A \otimes \overline{A}^*,b)$ is the <u>normalized Hochschild complex</u>.

<u>Notation</u>: $(a_o;a_1,\dots,a_n) = (a_o,a_1,\dots,a_n) \bmod D_n$

(in $A \otimes \overline{A}^n = A^{n+1}/D_n$)

$(a_o;a_1,\dots,a_n) = 0$ whenever one of the a_i, $1 \leq i \leq n$, lies in $\mathrm{Im}(k \to A)$

(3) $H_n(A) = H_n(A \otimes \overline{A}^*,b)$, $n \geq 0$.

This is an immediate consequence of the acyclicity of (D_*,b): look at the long exact homology sequence of

$$0 \longrightarrow (D_*,b) \longrightarrow (A^{*+1},b) \longrightarrow (A \otimes \bar{A}^*,b) \longrightarrow 0$$

In other words: the projection $(A^{*+1},b) \to (A \otimes \bar{A}^*,b)$ is a quasi-isomorphism.

(4) A final remark concerning the induced boundary operator

$b: A \otimes \bar{A}^n \to A \otimes \bar{A}^{n-1}$.

We have explicitely:

$$b(a_o;a_1,\ldots,a_n) = (a_oa_1;a_2,\ldots,a_n) - (a_o;a_1a_2,\ldots,a_n) + \ldots$$

$$+ (-1)^n(a_na_o;a_1,\ldots,a_{n-1})$$

All products involved have to be interpreted as follows: Multiply any representatives in A and reduce "mod k" (for the middle terms). Everything is well-defined by the fact that (D_*,b) is a subcomplex of (A^{*+1},b). In most applications $A = k \oplus \bar{A}$, an augmented k-algebra, and there is no deeper problem about the notation.

Remark-Definition 2.5.2 Consider now $C(A) = (A^{*+1},b,B)$, the mixed Hochschild complex of A, and recall the explicit formula for B: $A^{n+1} \to A^{n+2}$:

$$B(a_o,\ldots,a_n) = \sum_{i=0}^{n} (-1)^{in}(1,a_i,\ldots,a_n,a_o,\ldots,a_{i-1})$$

$$+ \sum_{i=0}^{n} (-1)^{in}(a_i,1,a_{i+1},\ldots,a_{i-1})$$

This shows that $BD_n \subset D_{n+1}$.

Hence we have the mixed subcomplex (D_*,b,B) of $C(A) = (A^{*+1},b,B)$, which yields an exact sequence of mixed complexes

$$0 \longrightarrow (D_*,b,B) \longrightarrow (A^{*+1},b,B) \longrightarrow (A \otimes \bar{A}^*,b,B) \longrightarrow 0$$

where $\bar{C}(A) = (A \otimes \bar{A}^*,b,B)$ is the normalized mixed Hochschild complex of A.

The explicit formula for $B: A \otimes \bar{A}^n \to A \otimes \bar{A}^{n+1}$ reads now like this:

$$B(a_o;a_1,\ldots,a_n) = \sum_{i=0}^{n} (-1)^{in}(1;a_i,a_{i+1},a_n,a_o,\ldots,a_{i-1})$$

Proposition 2.5.3 The projection $C(A) \to \overline{C}(A)$ of the mixed Hochschild complex on the normalized mixed Hochschild complex is a quasi-isomorphism (i.e. induces isomorphisms both in homology and in cyclic homology).

Proof. Consider the commutative diagram

$$\begin{array}{ccccccccc} 0 & \longrightarrow & (C(A),b) & \longrightarrow & ({}_BC(A),d) & \longrightarrow & ({}_BC(A)[2],d[2]) & \longrightarrow & 0 \\ & & \downarrow & & \downarrow & & \downarrow & & \\ 0 & \longrightarrow & (\overline{C}(A),b) & \longrightarrow & ({}_B\overline{C}(A),d) & \longrightarrow & ({}_B\overline{C}(A)[2],d[2]) & \longrightarrow & 0 \end{array}$$

and apply 2.3.15 together with 2.5.1(3).

Remark 2.5.4 We have until yet the following indentifications:

$H_*(A) = H_*(C(A)) = H_*(\overline{C}(A))$ (homology)

$HC_*(A) = HC_*(C(A)) = HC_*(\overline{C}(A))$ (cyclic homology)

Hypothesis 2.5.5 For all the rest of this paragraph we shall assume that k is a k-direct summand of A (in particular: $k \to A$ is injective). Note that A need not be augmented (think of Clifford algebras for example).

Remark 2.5.6 The inclusion $k \to A$ induces a homomorphism of mixed complexes $\overline{C}(k) \to \overline{C}(A)$, which reads as a chain transformation like this:

$$\begin{array}{ccc} \overline{C}(k) & \longrightarrow & \overline{C}(A) \\ \vdots & & \vdots \\ 0 & \longrightarrow & A \otimes \overline{A}^3 \\ \downarrow & & \downarrow b \\ 0 & \longrightarrow & A \otimes \overline{A}^2 \\ \downarrow & & \downarrow b \\ 0 & \longrightarrow & A \otimes A \\ \downarrow & & \downarrow \\ k & \longrightarrow & A \end{array}$$

The mixed quotient complex $(\overline{C}(A)_{red},b,B)$ is given by

$$\overline{C}_o(A)_{red} = \overline{A}$$

$$\overline{C}_n(A)_{red} = \overline{C}_n(A) = A \otimes \overline{A}^n, \quad n \geq 1$$

and

$$b: A \otimes \overline{A} \longrightarrow \overline{A}$$

$$(a_o;a_1) \longrightarrow a_o a_1 \quad \text{mod } k$$

$$B: \overline{A} \longrightarrow A \otimes \overline{A}$$

$$a \longrightarrow (1;a)$$

We have an exact sequence of mixed complexes

$$0 \longrightarrow \overline{C}(k) \longrightarrow \overline{C}(A) \longrightarrow \overline{C}(A)_{red} \longrightarrow 0$$

(which is only nontrivial in degree 0).

Definition 2.5.7 In the situation 2.5.5 define reduced Hochschild homology and reduced cyclic homology of A by

$$\overline{H}_*(A) = H_*(\overline{C}(A)_{red})$$

$$\overline{HC}_*(A) = HC_*(\overline{C}(A)_{red}).$$

Remark 2.5.8 Relation between Hochschild homology and reduced Hochschild homology.
We have
(i) an exact sequence

$$0 \longrightarrow H_1(A) \longrightarrow \overline{H}_1(A) \longrightarrow k \longrightarrow H_o(A) \longrightarrow \overline{H}_o(A) \longrightarrow 0$$

(ii) $\overline{H}_n(A) = H_n(A)$ for $n \geq 2$.

The long exact homology sequence associated with our exact sequence of chain complexes

$$0 \longrightarrow (\overline{C}(k),b) \longrightarrow (\overline{C}(A),b) \longrightarrow (\overline{C}(A)_{red},b) \longrightarrow 0$$

corresponds in lowest degrees to (i), in higher degrees to (ii).

<u>Proposition 2.5.9</u> There are two long exact sequences

$$(1) \longrightarrow HC_n(k) \longrightarrow HC_n(A) \longrightarrow \overline{HC}_n(A) \longrightarrow HC_{n-1}(k) \longrightarrow$$

$$(2) \longrightarrow \overline{H}_n(A) \xrightarrow{I} \overline{HC}_n(A) \xrightarrow{S} \overline{HC}_{n-2} \xrightarrow{B} \overline{H}_{n-1}(A) \longrightarrow$$

<u>Proof</u>. (1) is the long exact homology sequence of the exact sequence of chain complexes

$$0 \longrightarrow ({}_B\overline{C}(k),d) \longrightarrow ({}_B\overline{C}(A),d) \longrightarrow ({}_B\overline{C}(A)_{red},d) \longrightarrow 0$$

(note that ${}_B\overline{C}(k)_n = \begin{cases} k & n \text{ even} \\ 0 & n \text{ odd} \end{cases}$ in correspondence with the appearance or non-appearance of the A- and $\overline{A}$-term in ${}_B\overline{C}(A)_n$ and $({}_B\overline{C}(A)_{red})_n$ for even or odd n).

(2) is the long exact homology sequence of

$$0 \longrightarrow (\overline{C}(A)_{red},b) \longrightarrow ({}_B\overline{C}(A)_{red},d) \longrightarrow ({}_B\overline{C}(A)_{red}[2],d[2]) \longrightarrow 0$$

<u>Remark 2.5.10</u>

(a) The first long exact sequence of 2.5.9 actually splits in 5-term exact sequences

$$0 \longrightarrow HC_{2m+1}(A) \longrightarrow \overline{HC}_{2m+1}(A) \longrightarrow k \longrightarrow HC_{2m}(A) \longrightarrow \overline{HC}_{2m}(A) \longrightarrow 0, \; m \geq 0.$$

(b) The second long exact sequence of 2.5.9 yields in lowest degrees (cf. 2.3.7)

(i) an isomorphism $\overline{H}_o(A) = \overline{HC}_o(A)$

(ii) an epimorphism $\overline{H}_1(A) \to \overline{HC}_1(A) \to 0$

<u>Example 2.5.11</u> Let k be a commutative noetherian ring, $A = M_r(k)$ the k-algebra of r×r-matrices with coefficients in k.

Then $\overline{HC}_n(A) = 0$ for all $n \geq 0$.

(i) We first look at reduced Hochschild homology $\overline{H}_*(A)$. I claim that $\overline{H}_*(A) = 0$ (in all degrees).

$H_n(A) = 0$ for $n \geq 1$ (cf. 2.4.8), hence

$\overline{H}_n(A) = 0$ for $n \geq 2$ (2.5.8)

It remains to show that $\overline{H}_o(A) = \overline{H}_1(A) = 0$

Write $A = k1 + \overline{A}$ (k-direct sum)

where $\overline{A} = \sum_{i=1}^{r-1} k(1-e_{ii}) + \sum_{i\neq j} ke_{ij}$ (relative to the standard k-basis of A)

Now, $\overline{A} = [A,A]$, and thus $\overline{H}_o(A) = \overline{A}/\overline{[A,A]} = 0$.

The five-term exact sequence 2.5.8 (i) becomes

$$0 \longrightarrow \overline{H}_1(A) \longrightarrow k \longrightarrow H_o(A) \longrightarrow 0$$

and $H_o(A) = A/[A,A] = k$.

Since k is noetherian, the surjective k-endomorphism $k \to k = H_o(A)$ most be injective too.

Finally $\overline{H}_1(A) = 0$.

(ii) We now pass to reduced cyclic homology.

By 2.5.10(b) we have $\overline{HC}_o(A) = \overline{HC}_1(A) = 0$.

Since $\overline{H}_n(A) = 0$ for $n \geq 1$, we obtain by 2.5.9 (2) (as in 2.4.8):

$\overline{HC}_n(A) = \overline{HC}_{n-2}(A)$ for $n \geq 2$. This proves our assertion.

Remark 2.5.12 Assume now that $A = k \oplus \overline{A}$ is an augmented k-algebra. The commutative diagram

$$\begin{array}{ccc} k & \longrightarrow & A \\ \downarrow \text{id} & \swarrow & \\ k & & \end{array}$$

of k-algebra homomorphisms gives rise to a commutative diagram of mixed complexes

$$\begin{array}{ccccc} 0 & \longrightarrow & \overline{C}(k) & \longrightarrow & \overline{C}(A) \\ & & \downarrow \text{id} & \swarrow & \\ & & \overline{C}(k) & & \end{array}$$

i.e. to a splitting of the exact sequence of chain complexes

$$0 \longrightarrow {}_B\overline{C}(k) \longrightarrow {}_B\overline{C}(A) \longrightarrow {}_B\overline{C}(A)_{red} \longrightarrow 0.$$

Hence the long exact sequence 2.5.9 (1) splits too, i.e. we have

$$HC_*(A) = HC_*(k) \oplus \overline{H}C_*(A)$$

More explicitely:

$$HC_n(A) = \begin{cases} k \oplus \overline{H}C_n(A) & n \text{ even} \\ \overline{H}C_n(A) & n \text{ odd} \end{cases}$$

Note that in example 2.5.11 we obtained the same result in a non-augmented setting.

Example 2.5.13 Cyclic homology of a tensor algebra. Let V be a flat k-module, $A = T(V) = \bigoplus_{m\geq 0} V^m$ the tensor algebra of V over k. (V^m means m-fold tensor product over k)

(1) Hochschild homology of $A = T(V)$.

(a) The acyclic Hochschild complex (A^{*+1}, b') gives rise to an exact sequence of A-A-bimodules

$$0 \longrightarrow A \otimes V \otimes A \xrightarrow{b'} A \otimes A \xrightarrow{b'} A \longrightarrow 0$$

Recall ([C.E., p.168]) that $J = \mathrm{Ker}(A \otimes A \overset{b'}{\to} A)$ is the module of non-commutative differentials of A:
For every A-A-bimodule M and for every derivation $d: A \to M$ there is a unique factorization

$$\begin{array}{ccc} A & \xrightarrow{d} & M \\ \downarrow j & \nearrow f & \\ J & & \end{array}$$

where $j(a) = a \otimes 1 - 1 \otimes a$, $a \in A$, and where f is an A-A-bimodule homomorphism.
Since derivations on $A = T(V)$ are uniquely definable and determined by their (k-linear) restrictions on V, the A-A-bimodule $A \otimes V \otimes A$ has the same universal factorization property. Thus $A \otimes V \otimes A \underset{b'}{\simeq} J$.

(b) The Hochschild homology of $A = T(V)$ is given by

$$H_0(A) = \bigoplus_{m\geq 0} V^m/(1-\sigma), \quad H_1(A) = \bigoplus_{m\geq 1} (V^m)^\sigma, \quad H_n(A) = 0 \quad \text{for } n \geq 2,$$

where $\sigma : V^m \to V^m$ is the cyclic permutation

$\sigma(v_1,\ldots,v_m) = (v_m,v_1,\ldots,v_{m-1})$.

Proof. Consider the long exact homology sequence

$$\ldots \to \mathrm{Tor}_n^{A^e}(A, A\otimes V\otimes A) \longrightarrow \mathrm{Tor}_n^{A^e}(A,A^e) \longrightarrow \mathrm{Tor}_n^{A^e}(A,A) \to \ldots$$

$$\ldots \to \mathrm{Tor}_o^{A^e}(A, A\otimes V\otimes A) \longrightarrow \mathrm{Tor}_o^{A^e}(A,A^e) \longrightarrow \mathrm{Tor}_o^{A^e}(A,A) \to 0$$

Since A^e and $A\otimes V\otimes A$ are A^e-flat, we obtain

(i) $H_n(A) = \mathrm{Tor}_n^{A^e}(A,A) = 0$ for $n \geq 2$

(ii) an exact sequence in lowest degrees:

$$\begin{array}{ccccccccc} 0 \longrightarrow H_1(A) \longrightarrow & A\otimes_{A^e}(A\otimes V\otimes A) & \xrightarrow{1\otimes b'} & A\otimes A^e & \longrightarrow H_o(A) \longrightarrow 0 \\ & \| & & \| & \\ & A\otimes V & \xrightarrow{\quad b \quad} & A & \end{array}$$

Recall: $b\colon A\otimes A \to A$ is given by $b(a_o,a_1) = a_oa_1 - a_1a_o$.
Spezializing $a_o = (v_1,\ldots,v_m) \in V^{m-1}$, $a_1 = v_m \in V$:

$$b((v_1,\ldots,v_{m-1})\otimes v_m) = (v_1,\ldots,v_m) - (v_m,v_1,\ldots,v_{m-1}) = (1-\sigma)(v_1,\ldots,v_m)$$

We obtain finally:

$$H_o(A) = \mathrm{CoKer}(A\otimes V \xrightarrow{b} A) = \bigoplus_{m\leq 0} V^m/(1-\sigma)$$

$$H_1(A) = \mathrm{Ker}(A\otimes V \xrightarrow{b} A) = \bigoplus_{m\leq 1} (V^m)^\sigma$$

(c) Reduced Hochschild homology of $A = T(V)$.
We have (2.5.8):
(i) an exact sequence

$$0 \to H_1(A) \to \overline{H}_1(A) \to k \to H_o(A) \to \overline{H}_o(A) \to 0$$

(ii) $\overline{H}_n(A) = H_n(A) = 0$, $n \geq 2$

Now, $[A,A] \subset \overline{A} = \bigoplus_{m\geq 1} V^m$, and hence

$$\begin{array}{ccccccccc}
0 & \longrightarrow & k & \longrightarrow & H_0(A) & \longrightarrow & \overline{H}_0(A) & \longrightarrow & 0 \\
 & & & & \| & & \| & & \\
0 & \longrightarrow & k & \longrightarrow & \bigoplus_{m\geq 0} V^m/(1-\sigma) & \longrightarrow & \bigoplus_{m\geq 1} V^m(1-\sigma) & \longrightarrow & 0
\end{array}$$

is exact.

This implies $\overline{H}_1(A) = H_1(A) = \bigoplus_{m\geq 1} (V^m)^\sigma$.

(2) (Reduced) cyclic homology of $A = T(V)$.

(a) Identification of $B: \overline{HC}_0(A) \to \overline{H}_1(A)$

We have $\overline{HC}_0(A) = \overline{H}_0(A)$ (2.5.10)

and $\overline{H}_1(A) = H_1(A)$.

B is induced by $B: \overline{A} \longrightarrow A \otimes \overline{A}$

$a \mapsto (1;a)$

<u>Lemma</u>: The following square is commutative

$$\begin{array}{ccc}
\overline{HC}_0(A) & \xrightarrow{B} & \overline{H}_1(A) \\
\| & & \| \\
\bigoplus_{m\geq 1} V^m/(1-\sigma) & \xrightarrow{\oplus \nu_m} & \bigoplus_{m\geq 1} (V^m)^\sigma
\end{array}$$

where $\nu_m = \sum_{i=0}^{m-1} \sigma^i$ (norm map).

(Note that we are dealing with operations of the various cyclic groups <u>inside</u> $A = T(V)$. Don't confound with the operations on the Hochschild complexes of A)

<u>Proof</u>. Note that $(x;yz) \equiv (xy;z) + (zx;y)$ in

$\overline{H}_1(A) = \mathrm{Ker}(A \otimes \overline{A} \xrightarrow{b} \overline{A})/b(A \otimes \overline{A}^2)$.

Consider $a = v_1 \dots v_m \in V^m \subset A$.

$B(a) = (1;a) \bmod b(A \otimes \overline{A}^2)$

But (recall (1)(b)):

$$(1;v_1\dots v_m) \equiv (v_1;v_2\dots v_m) + (v_2\dots v_m;v_1)$$

$$\equiv (v_1v_2;v_3\dots v_m) + (v_3\dots v_mv_1;v_2) + (v_2\dots v_m;v_1)$$

$$\equiv \sum_{i=1}^{m} (v_{i+1} \dots v_{i-1} ; v_i) \in \overline{H}_1(A) \subset A \otimes V$$

Identifying $V^{m-1} \otimes V \subset A \otimes V$ with $V^m \subset A$, we obtain $(1;v_1 \dots v_m) = \nu_m(v_1 \dots v_m)$ as claimed.

(b) Recall the long exact sequence 2.5.9 (2) relating reduced Hochschild homology and reduced cyclic homology. We obtain in our special case

(i) $\overline{H}_o(A) = \overline{HC}_o(A)$ (always true)

(ii) a four-term exact sequence

$$0 \longrightarrow \overline{HC}_2(A) \longrightarrow \overline{HC}_o(A) \xrightarrow{B} \overline{H}_1(A) \longrightarrow \overline{HC}_1(A) \longrightarrow 0$$

(iii) isomorphisms $\overline{HC}_n(A) \simeq \overline{HC}_{n-2}(A)$, $n \geq 3$.

Conclusion: $\overline{HC}_2(A) \simeq \operatorname{Ker} B$

$\overline{HC}_1(A) \simeq \operatorname{CoKer} B$

and consequently

$$\overline{HC}_n(A) \simeq \begin{cases} \operatorname{Ker}(\overline{HC}_o(A) \xrightarrow{B} \overline{H}_1(A)) & n \text{ even} \\ \operatorname{CoKer}(\overline{HC}_o(A) \xrightarrow{B} \overline{H}_1(A)) & n \text{ odd} \end{cases} \qquad n \geq 1$$

On the other hand, we have the following

<u>Lemma</u>: Let $G_m = \langle \sigma \rangle$ be a finite cyclic group of order m, M a G_m-module, $\nu: M \to M$ the norm operator $\nu = \sum_{i=0}^{m-1} \sigma^i$.

Then $H_o(G_m, M) = M/(1-\sigma)M$

$$H_n(G_m, M) = \begin{cases} \operatorname{Ker} \nu / \operatorname{Im}(1-\sigma) & n \text{ even} \\ \operatorname{Ker}(1-\sigma)/\operatorname{Im} \nu & n \text{ odd} \end{cases} \qquad n \geq 1$$

<u>Proof</u>. [H.St., p.201].

This lemma, together with the above lemma identifying the operator B as a sum of norm maps, give the

<u>Proposition</u>: $\overline{HC}_n(T(V)) = \bigoplus_{m \geq 1} H_n(G_m, V^m)$, $n \geq 0$

where the cyclic group G_m acts on V^m via σ.

Proof. $\mathrm{Ker}(1-\sigma) = (V^m)^\sigma$. Put the two lemmas together.

(c) $HC_n(T(V)) = HC_n(k) \oplus \overline{HC}_n(T(V))$ by 2.5.12.

Assume now that $\mathbb{Q} \subset k$. Then

$$\overline{HC}_o(T(V)) = \bigoplus_{m\geq 1} V^m/(1-\sigma)$$

$$\overline{HC}_n(T(V)) = 0 \quad \text{for } n \geq 1.$$

and consequently

$$HC_o(T(V)) = \bigoplus_{m\geq 0} V^m/(1-\sigma)$$

$$HC_n(T(V)) = \begin{cases} k & n \text{ even} \\ 0 & n \text{ odd} \end{cases} \qquad n \geq 1$$

Remark 2.5.14 Cyclic homology of non-unital associative k-algebras.

A short inspection of the definitions 2.1.4 and 2.2.1 (the double complex $C(A)$ and $HC_*(A) = H_*(\mathrm{Tot}(C(A)))$) shows that these definitions make sense for any associative k-algebra A, unital or not.

(Caution: There is no longer a contracting homotopy s for the odd degree columns of $C(A)$).

Furthermore, 2.2.2 (flat extensions of scalars) and 2.2.3 (direct limits) remain valid in the non-unital setting.
In particular, let A be a unital associative k-algebra, $M(A) = \varinjlim M_r(A)$ the algebra of infinite matrices with only a finite number of nonzero entries in A, then $M(A)$ is non-unital, but nevertheless

$HC_*(M(A)) = \varinjlim HC_*(M_r(A))$ (= $HC_*(A)$: Morita-invariance of cyclic homology; cf. 2.7.14).

At this stage, we only can prove that

$$HC_*(M(k)) = \varinjlim HC_*(M_r(k)) = HC_*(k) \quad (2.4.8)$$

The connection between non-unital cyclic homology and reduced unital cyclic homology is simple:

Proposition 2.5.15 Let $A = k \oplus I$ be an augmented k-algebra. Then

$HC_*(I) = \overline{H}C_*(A)$.

Proof. We want to establish an isomorphism of chain complexes

$$h\colon T = \mathrm{Tot}(C(I)) \to {}_B\overline{C}(A)_{red}$$

which will immediately give our result.
For $r \geq 1$ consider the ismorphisms

$$I^{r+1} \oplus I^r \longrightarrow A \otimes I^r$$

$$((x_o,\ldots,x_r),(y_1,\ldots,y_r)) \to (x_o;x_1,\ldots,x_r) + (1;y_1,\ldots,y_r)$$

which yield isomorphisms of graded modules

$$\begin{array}{ccc} T_n & = & I^{n+1} \oplus I^n \oplus I^{n-1} \oplus I^{n-2} \oplus \ldots \\ \downarrow h_n & & \wr| \qquad\qquad \wr| \\ ({}_B\overline{C}(A)_{red})_n & = & (A \otimes I^n) \oplus \quad (A \otimes I^{n-2}) \oplus \ldots \end{array}$$

(Note that for even n the last component isomorphism is given by the identity on $I = \overline{A}$)

It remains to verify that the h_n commute with the differentials $d = b \oplus (D-b') \oplus N + b \oplus ..$ on T and $d = b + B$ on ${}_B\overline{C}(A)_{red}$.
This follows from the identities

(1) $bh(x_o,\ldots,x_r) = b(x_o;x_1,\ldots,x_r) = hb(x_o,\ldots,x_r)$

(both b are given by the same formula, "up to a semicolon")

(2) $Bh(y_1,\ldots,y_r) = B(1;y_1,\ldots,y_r) = 0$

(3) $bh(y_1,\ldots,y_r) = b(1;y_1,\ldots,y_r)$

$$= (y_1;y_2,\ldots,y_r) + (-1)^r(y_r;y_1,\ldots,y_{r-1})$$

$$+ \sum_{i=1}^{r-1} (-1)^i (1;y_1,\ldots,y_iy_{i+1},\ldots,y_r)$$

$$= h((D-b')(y_1,\ldots,y_r)$$

(4) $Bh(x_o,x_1,\dots,x_r) = B(x_o;x_1,\dots,x_r)$

$$= \sum_{i=0}^{r} (-1)^{ir}(1;x_i,\dots,x_r,x_o,\dots,x_{i-1})$$

$$= h(N(x_o,\dots,x_r))$$

Example 2.5.16 $A = k \oplus I$, a ring of dual numbers (i.e. $I^2 = 0$). In this particular situation, the differentials b and b' on $C(I)$ are zero.

Thus: $\overline{HC}_n(A) = HC_n(I) = \bigoplus_{K=0}^{n} H_{n-K}(G_{K+1}, I^{K+1})$

(group homology)

Assume now that $\mathbb{Q} \subset k$.

We obtain $\overline{HC}_n(A) = I^{n+1}/(1-t)I^{n+1}$ (cf. lemma in the proof of 2.2.6)

Spezializing to $A = k[\varepsilon]$, the usual ring of dual numbers over k, this gives

$$HC_n(k[\varepsilon]) = \begin{cases} k \oplus k & n \text{ even} \\ 0 & n \text{ odd} \end{cases}$$

($I^{n+1} = k$ for all $n \geq 0$, and $t.1 = 1$ for n even, $t.1 = -1$ for n odd)

Remark 2.5.17 Let A be a unital associative k-algebra, $A = k \oplus \overline{A}$ as a k-module (with $\overline{A} = \text{CoKer}(k \to A)$).

(1) Write $A \otimes \overline{A}^r = \overline{A} \otimes \overline{A}^r \oplus 1 \otimes \overline{A}^r$ (k-direct sum), where $1 \otimes \overline{A}^r$ is the k-submodule of $A \otimes \overline{A}^r$ spanned by the elements $(1;a_1,\dots,a_r)$. We have (as in the proof of 2.5.15) ismorphisms

$$\overline{A}^{r+1} \oplus \overline{A}^r \xrightarrow{\quad h \quad} A \otimes \overline{A}^r$$

$$((x_o,\dots,x_r),(y_1,\dots,y_r)) \to (x_o;x_1,\dots,x_r) + (1;y_1,\dots,y_r)$$

Now, the operators $D,N\colon \overline{A}^{r+1} \to \overline{A}^{r+1}$ make sense, and the formulas (3) and (4) in the proof of 2.5.15 read:

(3) $h(D(y_1,\dots,y_{r+1})) = bh(y_1,\dots,y_{r+1}) \bmod 1 \otimes \overline{A}^r$

(4) $h(N(x_o,\ldots,x_r)) = B(h(x_o,\ldots,x_r))$

We thus obtain the following commutative diagram (note that $B(1 \otimes \bar{A}^r) = 0$):

$$\begin{array}{ccccc}
A \otimes \bar{A}^r/1 \otimes \bar{A}^r & \xleftarrow{b} & 1 \otimes \bar{A}^{r+1} & \xleftarrow{B} & A \otimes \bar{A}^r/1 \otimes \bar{A}^r \\
\| h & & \| h & & \| h \\
\bar{A}^{r+1} & \xleftarrow{D} & \bar{A}^{r+1} & \xleftarrow{N} & \bar{A}^{r+1}
\end{array}$$

(2) We have an exact sequence of k-modules

$$1 \otimes A^n \to A^{n+1}/(1-t)A^{n+1} \to \bar{A}^{n+1}/(1-t)\bar{A}^{n+1} \to 0$$

(since $1 \otimes A^n \bmod(1-t) = (1 \otimes A^n + D_n)\bmod(1-t)$: for $(a_o,\ldots,a_n) \in D_n$ (2.5.1) there is K: $1 \leq K \leq n$ such that $t^K(a_o,\ldots,a_n) \in 1 \otimes A^n$, i.e. $(a_o,\ldots,a_n) = t^K(a_o,\ldots,a_n) + (1-t^K)(a_o,\ldots,a_n) \in 1 \otimes A^n + (1-t)A^{n+1}$)

Furthermore, $b(1,a_1,\ldots,a_n) = (1-t)(a_1,\ldots,a_n) \bmod 1 \otimes A^{n-1}$, i.e. the kernel of $A^{*+1}/(1-t) \to \bar{A}^{*+1}/(1-t)$ is a subcomplex of $(A^{*+1}/(1-t),b)$. Hence we have the chain complex $(\bar{A}^{*+1}/(1-t),b)$ (with induced differential b).

<u>Proposition 2.5.18</u> Assume that $\mathbb{Q} \subset k$, and that k is a k-direct summand in A. Then the complexes $(\bar{A}^{*+1}/(1-t),b)$ and $({}_B\bar{C}(A)_{red},d)$ are quasi-isomorphic, i.e. we have

$$\overline{HC}_*(A) = H_*(\bar{A}^{*+1}/(1-t),b).$$

<u>Proof</u>.

(1) Definition of a chain transformation

$$\pi: ({}_B\bar{C}(A)_{red},d) \to (\bar{A}^{*+1}/(1-t),b)$$

(which will reveal to be a quasi-isomorphism):

$$\begin{array}{ccc}
(\bar{A}^{*+1}/(1-t),b) & \xleftarrow{\pi} & ({}_B\bar{C}(A)_{red},d) \\
\vdots & & \vdots
\end{array}$$

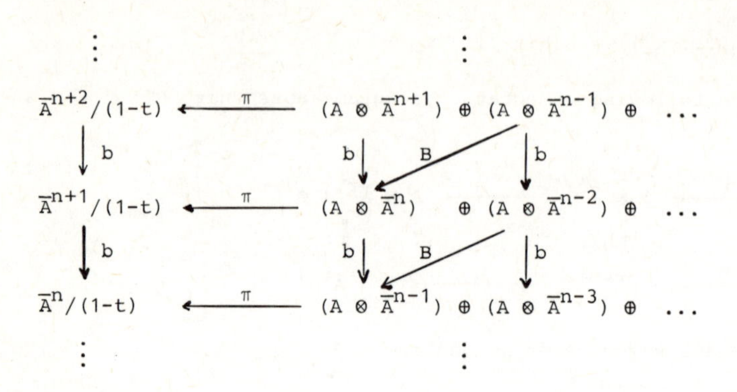

Explicitely: In degree n, π is a composition of the natural surjections

$$({}_B\overline{C}(A)_{red})_n \longrightarrow (A \otimes \overline{A}^n) \longrightarrow \overline{A}^{n+1} \longrightarrow \overline{A}^{n+1}/(1-t)$$

π is a homomorphism of chain complexes, since $B(A \otimes \overline{A}^{n-1}) \subset 1 \otimes \overline{A}^n$ (the k-submodule of $A \otimes \overline{A}^n$ spanned by the elements $(1;a_1,\dots,a_n)$) and thus $\pi \circ B = 0$.

(2) Definition of filtrations on ${}_B\overline{C}(A)_{red}$ and on $\overline{A}^{*+1}/(1-t)$ such that π becomes a morphism of filtered chain complexes

$$(F^p(\overline{A}^{*+1}/(1-t)))_n = \begin{cases} \overline{A}^{n+1}/(1-t) & n \le p \\ 0 & n > p \end{cases}$$

$$(F^p{}_B\overline{C}(A)_{red})_n = \begin{cases} ({}_B\overline{C}(A)_{red})_n & n \le p \\ (1 \otimes \overline{A}^{p+1}) + \sum_{K\ge 0} (A \otimes \overline{A}^{p-2K-1}) & n-p \text{ odd} \\ \sum_{K\ge 0} (A \otimes \overline{A}^{p-2K}) & n-p \text{ even} \end{cases} \quad (n > p \text{ for the last two})$$

Let us draw a picture of the couple of subcomplexes

$F^n(\overline{A}^{*+1}/(1-t))$ and $F^n({}_B\overline{C}(A)_{red})$

$$
\begin{array}{ccccccc}
0 & \longleftarrow & 0 & \oplus & (1 \otimes \overline{A}^{n+1}) & \oplus & (A \otimes \overline{A}^{n-1}) \\
\downarrow & & \downarrow & & \downarrow b \quad {\scriptstyle B}\swarrow & & \downarrow b \\
0 & \longleftarrow & 0 & \oplus & (A \otimes \overline{A}^{n}) & \oplus & (A \otimes \overline{A}^{n-2}) \\
\downarrow & & \downarrow \quad {\scriptstyle B}\swarrow & & \downarrow b \quad {\scriptstyle B}\swarrow & & \downarrow b \\
0 & \longleftarrow & (1 \otimes \overline{A}^{n+1}) & \oplus & (A \otimes \overline{A}^{n-1}) & \oplus & (A \otimes \overline{A}^{n-3}) \\
\downarrow & & \downarrow b \quad {\scriptstyle B}\swarrow & & \downarrow b \quad {\scriptstyle B}\swarrow & & \downarrow b \\
\overline{A}^{n+1}/(1-t) & \xleftarrow{\pi} & (A \otimes \overline{A}^{n}) & \oplus & (A \otimes \overline{A}^{n-2}) & \oplus & (A \otimes \overline{A}^{n-4}) \\
\downarrow b & & \downarrow b \quad {\scriptstyle B}\swarrow & & \downarrow b \quad {\scriptstyle B}\swarrow & & \downarrow b \\
\overline{A}^{n}/(1-t) & \xleftarrow{\pi} & (A \otimes \overline{A}^{n-1}) & \oplus & (A \otimes \overline{A}^{n-3}) & \oplus & (A \otimes \overline{A}^{n-5})
\end{array}
$$

As $B(A \otimes \overline{A}^n) \subset a \otimes \overline{A}^{n+1}$, $F^n{}_B\overline{C}(A)_{red}$ is indeed a subcomplex of ${}_B\overline{C}(A)_{red}$.

(Convention: $A \otimes \overline{A}^0 = \overline{A}$ in the reduced case).

(3) Description of the quotient filtrations.

Write $\overline{A}$ for the filtered chain complex $(\overline{A}^{*+1}/(1-t),b)$ and $\overline{C}$ for the filtered chain complex $({}_B\overline{C}(A)_{red},d)$

When passing to the quotient filtrations, we obtain

$$
(F^p\overline{A}/F^{p-1}\overline{A})_{p+q} = \begin{cases} 0 & q < 0 \\ \overline{A}^{p+1}/(1-t) & q = 0 \\ 0 & q > 0 \end{cases}
$$

$$
(F^p\overline{C}/F^{p-1}\overline{C})_{p+q} = \begin{cases} 0 & q < 0 \\ A \otimes \overline{A}^p/1 \otimes \overline{A}^p & q = 0 \\ 1 \otimes \overline{A}^{p+1} \quad q \text{ odd} & q > 0 \\ A \otimes \overline{A}^p/1 \otimes \overline{A}^p \quad q \text{ even} & q > 0 \end{cases}
$$

We are interested in the differential of $F^p\overline{C}/F^{p-1}\overline{C}$.

Recall the commutative diagram of 2.5.17 (1):

$$\begin{array}{ccccc} A \otimes \bar{A}^p/1 \otimes \bar{A}^p & \xleftarrow{b} & 1 \otimes \bar{A}^{p+1} & \xleftarrow{B} & A \otimes \bar{A}^p/1 \otimes \bar{A}^p \\ \| h & & \| h & & \| h \\ \bar{A}^{p+1} & \xleftarrow{D} & \bar{A}^{p+1} & \xleftarrow{N} & \bar{A}^{p+1} \end{array}$$

Thus we obtain: Since $\mathbb{Q} \subset k$, $F^p\bar{C}/F^{p-1}\bar{C}$ is <u>acyclic</u> (in positive dimension) for every $p \geq 0$.

(4) π is a quasi-isomorphism:

$$E^1_{p,q}(\bar{C}) = H_{p+q}(F^p\bar{C}/F^{p-1}\bar{C}) = \begin{cases} \bar{A}^{p+1}/(1-t) & q = 0 \\ 0 & q \neq 0 \end{cases}$$

$$E^1_{p,q}(\bar{A}) = H_{p+q}(F^p\bar{A}/F^{p-1}\bar{A}) = \begin{cases} \bar{A}^{p+1}/(1-t) & q = 0 \\ 0 & q \neq 0 \end{cases}$$

π^1: $E^1(\bar{C}) \to E^1(\bar{A})$ is the identity.

By the approximation theorem 1.3.7 we obtain that $H_*(\pi)$: $H_*(\bar{C}) \to H_*(\bar{A})$ must be an isomorphism.

I.2.6 Cyclic Cohomology

<u>Remark 2.6.1</u> Let k be a commutative ring, A a unital associative k-algebra. Recall the definition of the double chain complex $C(A)$ (2.1.4).
Applying the functor $\mathrm{Hom}_k(-,k)$, we obtain the double cochain complex $C^t(A) = \mathrm{Hom}_k(C(A),k)$.

(We write $(\ )^t$ for "transposed" in order to avoid a notational $*$-invasion). Explicitely:

$$\begin{array}{lccccc} C^t(A)^{2,*} & \mathrm{Hom}_k(A^3,k) & \xrightarrow{D^t} & \mathrm{Hom}_k(A^3k) & \xrightarrow{N^t} & \mathrm{Hom}_k(A^3,k) \\ & \uparrow b^t & & \uparrow -b'^t & & \uparrow b^t \\ C^t(A)^{1,*} & \mathrm{Hom}_k(A^2,k) & \xrightarrow{D^t} & \mathrm{Hom}_k(A^2,k) & \xrightarrow{N^t} & \mathrm{Hom}_k(A^2,k) \\ & \uparrow b^t & & \uparrow -b'^t & & \uparrow b^t \\ C^t(A)^{0,*} & \mathrm{Hom}_k(A,k) & \xrightarrow{D^t} & \mathrm{Hom}_k(A,k) & \xrightarrow{N^t} & \mathrm{Hom}_k(A,k) \\ & & & & & \\ & C^t(A)^{*,0} & & C^t(A)^{*,1} & & C^t(A)^{*,2} \end{array}$$

$C(\ )$ is a covariant functor from the category of (unital) associative k-algebras to the category of double chain complexes over k (with morphisms of bidegree (0,0)), whereas $C^t(\ )$ is a contravariant functor from the category of (unital) associative k-algebras to the category of double cochain complexes over k.

(1) The (anti)commutativity of the relevant squares in $C^t(A)$ follows from 2.1.5 and the fact that $\mathrm{Hom}_k(-,k)$ is a contravariant functor.

(2) The explicit formulas:

$$b^t\tau(a_o,\dots,a_n) = (\tau\circ b)(a_o,\dots,a_n)$$

$$= \sum_{i=0}^{n-1}(-1)^i\tau(a_o,..,a_ia_{i+1},..,a_n) + (-1)^n\tau(a_na_o,a_1,..,a_{n-1})$$

$$b'^t\tau(a_o,..,a_n) = (\tau\circ b')(a_o,\dots,a_n)$$

$$= \sum_{i=0}^{n-1}(-1)^i\tau(a_o,\dots,a_ia_{i+1},\dots,a_n)$$

$$D^t\tau(a_o,\dots,a_n) = (\tau\circ D)(a_o,\dots,a_n)$$

$$= \tau(a_o,\dots,a_n) - (-1)^n\tau(a_n,a_o,\dots,a_{n-1})$$

$$N^t\tau(a_o,\dots,a_n) = (\tau\circ N)(a_o,\dots,a_n)$$

$$= \tau(a_o,\dots,a_n) + (-1)^n\tau(a_n,a_o,\dots,a_{n-1})$$

$$+ (-1)^{2n}\tau(a_{n-1},a_n,..,a_{n-2}) +..+(-1)^{n^2}\tau(a_1,a_2,..,a_n,a_o)$$

(3) The odd-degree columns $C^t(A)^{*,q}$, q odd, are acyclic, since the contracting homotopy operators dualizes, too.

(4) For the even-degree columns $C^t(A)^{*,q}$, q even, we obtain:

$${}'H^{p,q}(C^t(A)) = H^p(A,A^*)$$

(Hochschild cohomology of A with coefficients in $A^* = \mathrm{Hom}_k(A,k)$).

Whenever A is k-projective, then

$$H^p(A,A^*) = \mathrm{Ext}^p_{A^e}(A,A^*), \quad p \geq 0.$$

In this case, the acyclic Hochschild complex (A^{*+2},b'): $\dots \xrightarrow{b'} A^3 \xrightarrow{b'} A^2$

is an A^e-projective resolution of the A-A-bimodule A.
Note that our cochain complex $(\mathrm{Hom}_k(A^{*+1},k),b^t)$ identifies anyway with

$$(\mathrm{Hom}_{A^e}(A^e \underset{k}{\otimes} A^*, \mathrm{Hom}_k(A,k)), \mathrm{Hom}(b',1))$$

(cf. [C.E., pp.174/175]).

(5) Let us look now at the rows of $C^t(A)$. We get

$$''H^{p,q}(C^t(A)) = H^q(G_{p+1}, \mathrm{Hom}_k(A^{p+1},k))$$

(cohomology of the cyclic groups $G_{p+1} = \langle t \rangle$ of order $p+1$, $p \geq 0$, with coefficients in the G_{p+1}-modules $\mathrm{Hom}_k(A^{p+1},k)$: $(t\tau)(a_o,\ldots,a_p) = (-1)^p\tau(a_p,a_o,\ldots,a_{p-1})$).
Note that we have always

$$H^q(G_n,\mathrm{Hom}_k(A^n,k)) = \mathrm{Ext}^q_{\mathbb{Z}[G_n]}(\mathbb{Z},\mathrm{Hom}_k(A^n,k)), \quad n \geq 1:$$

Take the standard $\mathbb{Z}[G_n]$-free resolution of $\mathbb{Z}$:

$$\mathbb{Z} \xleftarrow{\varepsilon} \mathbb{Z}[G_n] \xleftarrow{D} \mathbb{Z}[G_n] \xleftarrow{N} \mathbb{Z}[G_n] \xleftarrow{D} \cdots$$

and apply $\mathrm{Hom}_{\mathbb{Z}[G_n]}(-, \mathrm{Hom}_k(A^n,k))$, $n \geq 1$. (cf. 2.1.3)

(6) We had the total chain complex $T = \mathrm{Tot}(C(A))$, associated with our double chain complex $C(A)$. When passing to the total cochain complex $\mathrm{Tot}(C^t(A))$ associated with the double cochain complex $C^t(A)$, we obtain precisely the dual complex T^t. Explicitely:

$$\begin{array}{ccccccc}
(T^t)^n & = (T_n)^t & = & \mathrm{Hom}_k(A^{n+1},k) & \oplus & \mathrm{Hom}_k(A^n,k) & \oplus \cdots \\
d_n^t \uparrow & & & \uparrow b^t+D^t & & \uparrow N^t-b'^t & \\
(T^t)^{n-1} & = (T_{n-1})^t & = & \mathrm{Hom}_k(A^n,k) & \oplus & \mathrm{Hom}_k(A^{n-1},k) & \oplus \cdots
\end{array}$$

<u>Definition 2.6.2</u> The <u>cyclic cohomology</u> $HC^*(A)$ of the unital associative k-algebra A is defined by

$$HC^n(A) = H^n(\mathrm{Tot}(C^t(A))), \quad n \geq 0.$$

Remark 2.6.3 This definition also makes sense in the non-unital case; you should only observe that the odd-degree columns of $C^t(A)$ now don't have a contracting homotopy (hence are not necessarily acyclic). This prevents from directly passing to comparison results relating Hochschild and cyclic cohomology in the non-unital setting. The remedy will be, as in the case of homology, reduced theory (cf. 2.5.14/15).

Proposition 2.6.4 Let k be a field, and let A be a (unital) associative k-algebra. Then

(a) $H^n(A,A^*) = (H_n(A))^*$

(b) $HC^n(A) = (HC_n(A))^*$

(k-dual), $n \geq 0$.

Proof. We need the following elementary

Lemma: Let k be a field, (X,d) a differential k-module and $(\mathrm{Hom}_k(X,k),d^t)$ its dual. Then the canonical homomorphism $H(\mathrm{Hom}_k(X,k), d^t) \to \mathrm{Hom}_k(H(X,d),k)$ given by $[u] \to ([x] \to ux)$ is an isomorphism. (see [Go, p.22]).

(a) Apply the lemma to $(X,d) = (A^{*+1},b)$.

(b) Apply the lemma to $(X,d) = (\mathrm{Tot}(C(A)),d)$ and recall 2.6.1 (6).

Remark 2.6.5 Note that 2.6.4 only depends on the exactness of $\mathrm{Hom}_k(-, k)$. If you consider (k,K)-cohomology in the following sense: $k \to K$ a homomorphism of commutative rings, A a unital associative k-algebra, $C^t_{(k,K)}(A) = \mathrm{Hom}_k(C(A),K)$, and $H^*_{(k,K)}(A,\mathrm{Hom}_k(A,K))$, $HC^*_{(k,K)}(A)$ defined via this modified double cochain complex, then 2.6.4 reads: if $k \to K$ is such that K is k-injective, the $HC^n_{(k,K)}(A) = \mathrm{Hom}_k(HC_n(A),K)$, $n \geq 0$.

Remark 2.6.6 (Flat extensions of scalars)
Let $k \to K$ be a flat homomorphism of commutative rings, A a unital associative k-algebra, ${}_KA = K \otimes_k A$ the K-algebra obtained by extension of scalars. Assume that A is a finitely presented k-module. Then

$HC^n({}_KA) = K \otimes HC^n(A)$, $n \geq 0$.

Proof. We have without any extra-assumption

$$\mathrm{Hom}_K(({}_KA)^n,K) = \mathrm{Hom}_K(K \underset{k}{\otimes} A^n,K) = \mathrm{Hom}_k(A^n,K)$$

But when K is k-flat and A finitely k-presented, then

$\mathrm{Hom}_k(A^n,K) = \mathrm{Hom}_k(A^n,k) \underset{k}{\otimes} K$ (cf. [Bou, A.X. 12]).

The rest of the proof is as in 2.2.2.

Note that in the spirit of the remark 2.6.5 we have always

$HC^n_{(k,K)}(A) = HC^n({}_KA)$, $n \geq 0$.

Remark 2.6.7 We have

$$"H^{n,o}(C^t(A)) = H^o(G_{n+1},\mathrm{Hom}_k(A^{n+1},k))$$

$$= \mathrm{Ker}(\mathrm{Hom}_k(A^{n+1},k) \xrightarrow{D^t} \mathrm{Hom}_k(A^{n+1},k))$$

$$= \{\tau \in \mathrm{Hom}_k(A^{n+1},k): \tau(a_o,..,a_n) = (-1)^n\tau(a_n,a_o,..,a_{n-1}), \underset{0\leq i\leq n}{a_i \in A}\}.$$

Moreover the formula $D^tb^t = b'^tD^t$ shows that $C^*_\lambda(A) = ("H^{*,o}(C^t(A)), b^t)$ is a subcomplex of the Hochschild cochain complex

$(C^t(A)^{*,o},b^t) = (\mathrm{Hom}_k(A^{*+1},k),b^t)$.

Put $H^n_\lambda(A) = H^n(C^*_\lambda(A))$, $n \geq 0$.

It is immediate that the map

σ: $C^*_\lambda(A) \rightarrow \mathrm{Tot}(C^t(A))$ given by

σ^n: $\mathrm{Ker}\ D^t \subset \mathrm{Hom}_k(A^{n+1},k) \subset (\mathrm{Tot}(C^t(A)))^n$, $n \geq 0$,

is a monomorphism of cochain complexes.

Proposition 2.6.8 Assume that $\mathbb{Q} \subset k$. Then $\sigma : C^*_\lambda(A) \rightarrow \mathrm{Tot}(C^t(A))$ is a quasi-isomorphism, i.e. $H^n(\sigma): H^n_\lambda(A) \rightarrow HC^n(A)$ is an isomorphism for every $n \geq 0$.

Proof. All arguments are dual to 2.2.6. Recall 1.4.12: Spectral sequence arguments apply to double cochain complexes of the first quadrant via visualization as double chain complexes of the third quadrant.
In order to fix notational ideas, let us consider the first filtration of $T^t = \mathrm{Tot}(C^t(A))$:

$$({}^{I}F_p T^t)^n = \bigoplus_{j\geq p} C^{j,n-j}$$

$$C^{p,q} = \mathrm{Hom}_k(C_{p,q},k) = \mathrm{Hom}_k(A^{p+1},k), \quad p \geq 0,\ q \geq 0.$$

We have:

$$H^n_\lambda(A) = {}'H^n{}''H^{n,0}(C^t(A)) = {}^{I}E_2^{n,0}, \quad \text{and}$$

$${}^{I}E_1^{p,q} = {}''H^{p,q}(C^t(A)) = H^q(G_{p+1},\mathrm{Hom}_k(A^{p+1},k)), \quad p,q \geq 0$$

Consequently:

$${}^{I}E_1^{p,q} = 0 \quad \text{for} \quad q > 0, \quad \text{since} \quad \mathbb{Q} \subset k$$

(once more (cf. [Ro, p.292]): G a finite group of order m, M an arbitrary G-module; then $mH^q(G,M) = 0$ for $q > 0$)
A fortiori: ${}^{I}E_2^{p,q} = 0$ for $q > 0$, and by 1.4.13 we obtain

$$H^n_\lambda(A) = {}^{I}E_2^{n,0} \simeq HC^n(A) = H^n(\mathrm{Tot}(C^t(A))), \quad n \geq 0.$$

Note that (as in 2.2.6) we should rigorously verify that our isomorphism is actually induced by σ. The argument is the same.

Convention 2.6.9 In order to avoid notational clumsiness we shall often suppress the $(\ )^t$-superscript: b will stand for b^t, b' for b'^t, s for s^t, $C(A)$ for $C^t(A)$, and so on.

Remark 2.6.10 We have now to formulate the mixed cochain complex approach to cohomology.

(1) A mixed cochain complex (M,b,B) is a non-negatively graded k-module $(M^n)_{n\geq 0}$ together with a degree +1 endomorphism b and a degree -1 endomorphism B such that $b^2 = B^2 = bB + Bb = 0$.
Thus (M,b) is cochain complex, (M,B) is a chain complex. Morphisms of mixed cochain complexes have to commute with both differentials.

Note that a priori mixed (chain) complexes and mixed cochain complexes are the same thing. There is only a different emphasis on what should be the primary differential and what should be the secondary differential.

(2) The associated cochain complex $({}^{B}M,d)$ of a mixed cochain complex (M,b,B) is defined by

$${}^{B}M^{n} = M^{n} \oplus M^{n-2} \oplus M^{n-4} \oplus \ldots$$

$$d^{n}(m^{n},m^{n-2},m^{n-4},\ldots) = (bm^{n}, bm^{n-2}+Bm^{n}, bm^{n-4}+Bm^{n-2}, \ldots)$$

(in short: $d = b + B$)

(3) We obtain the following exact sequence of cochain complexes

$$0 \longrightarrow ({}^{B}M[2],d[2]) \xrightarrow{S} ({}^{B}M,d) \xrightarrow{I} (M,b) \longrightarrow 0$$

which reads in degree n

$$0 \longrightarrow M^{n-2}\oplus M^{n-4} \oplus\ldots \xrightarrow{S} M^{n}\oplus M^{n-2}\oplus M^{n-4}\oplus\ldots \xrightarrow{I} M^{n} \longrightarrow 0$$

(S means injection, I means projection on the first factor).

(4) Let (M,b,B) be a mixed cochain complex.

$H^{*}(M) = H^{*}(M,b)$ the <u>cohomology of (M,b,B)</u>

$HC^{*}(M) = H^{*}({}^{B}M,d)$ the <u>cyclic cohomology of (M,b,B)</u>

(5) There is a long exact cohomology sequence

$$\ldots \longrightarrow H^{n}(M) \xrightarrow{B} HC^{n-1}(M) \xrightarrow{S} HC^{n+1}(M) \xrightarrow{I} H^{n+1}(M) \xrightarrow{B} \ldots$$

where the connecting homomorphism is induced by B. (this follows immediately from (3); cf. 2.3.6).
We have in lowest degrees:

(i) an isomorphism $HC^{0}(M) \xrightarrow{I} H^{0}(M)$

(ii) an exact sequence $0 \to HC^{1}(M) \xrightarrow{I} H^{1}(M) \xrightarrow{B} HC^{0}(M)$

(6) A morphism of mixed (cochain) complexes $F: (M,b,B) \to (N,b,B)$ gives rise to a commutative diagram of cochain complex homomorphisms

$$\begin{array}{ccccccccc}
0 & \longrightarrow & ({}^BM[2],d[2]) & \longrightarrow & ({}^BM,d) & \longrightarrow & (M,b) & \longrightarrow & 0 \\
 & & \downarrow {}^BF[2] & & \downarrow {}^BF & & \downarrow F & & \\
0 & \longrightarrow & ({}^BN[2],d[2]) & \longrightarrow & ({}^BN,d) & \longrightarrow & (N,b) & \longrightarrow & 0
\end{array}$$

and thus to a commutative diagram relating the long exact cohomology sequences

$$\begin{array}{ccccccccc}
\longrightarrow & H^n(M) & \longrightarrow & HC^{n-1}(M) & \longrightarrow & HC^{n+1}(M) & \longrightarrow & H^{n+1}(M) & \longrightarrow \\
 & \downarrow H^n(F) & & \downarrow & & \downarrow HC^{n+1}(F) & & \downarrow & \\
\longrightarrow & H^n(N) & \longrightarrow & HC^{n-1}(N) & \longrightarrow & HC^{n+1}(N) & \longrightarrow & H^{n+1}(N) & \longrightarrow
\end{array}$$

More generally, let $G^{(o)}: \ (M,b) \to (N,b)$ be a homomorphism of cochain complexes which allows a prolongation $G: ({}^BM,d) \quad ({}^BN,d)$ such that

$$\begin{array}{ccccccccc}
0 & \longrightarrow & ({}^BM[2],d[2]) & \longrightarrow & ({}^BM,d) & \longrightarrow & (M,b) & \longrightarrow & 0 \\
 & & \downarrow G[2] & & \downarrow G & & \downarrow G^{(o)} & & \\
0 & \longrightarrow & ({}^BN[2],d[2]) & \longrightarrow & ({}^BN,d) & \longrightarrow & (N,b) & \longrightarrow & 0
\end{array}$$

is commutative.

Then we have the same commutative diagram relating the long exact cohomology sequences (with $H^n(G^{(o)})$ and $HC^n(G)$ at the place of $H^n(F)$, $HC^n(F)$).

In this situation, the dual result to 2.3.15 is valid:

$G^{(o)}: (M,b) \to (N,b)$ is a quasi-isomorphism if and only if

$G: ({}^BM,d) \to ({}^BN,d)$ is a quasi-isomorphism.

<u>Definition 2.6.11</u> $C^t(A) = (\mathrm{Hom}_k(A^{*+1},k),b^t,B^t)$ is the <u>mixed (Hochschild) cochain complex</u> obtained by dualizing $C(A) = (A^{*+1},b,B)$. We shall write (cf. 2.6.9): $C(A) = (\mathrm{Hom}_k(A^{*+1},k),b,B)$.

<u>Remark 2.6.12</u>

(1) The cohomology $H^*(C(A))$ is the Hochschild cohomology $H^*(A,A^*)$ (by definition).

(2) The cyclic cohomology $HC^*(C(A))$ is the cyclic cohomology $HC^*(A)$ of A, as is shown by dualizing 2.4.4 and 2.4.5:
As $\mathrm{Tot}(C^t(A))$ is the k-dual of $\mathrm{Tot}(C(A))$, and ${}^BC(A)$ the k-dual of

${}_{B}C(A)$, we get immediately the homomorphism of cochain complexes f^t: $Tot(C^t(A)) \to {}^{B}C(A)$ given by $f^t = id + Ns$ which is a quasi-isomorphism (dualize the proof 2.4.5).

<u>Theorem 2.6.13</u> For every unital associative k-algebra A there is a long exact cohomology sequence

$$\longrightarrow H^n(A,A^*) \xrightarrow{B} HC^{n-1}(A) \xrightarrow{S} HC^{n+1}(A) \xrightarrow{I} H^{n+1}(A,A^*) \longrightarrow$$

<u>Proof</u>. 2.6.10(5) and 2.6.12.

<u>Complement 2.6.14</u> We have in lowest degrees

(i) an isomorphism $HC^o(A) \xrightarrow{I} H^o(A,A^*)$

(ii) a monomorphism $0 \to HC^1(A) \xrightarrow{I} H^1(A,A^*)$

<u>Example 2.6.15</u> Let k be a commutative ring , and let A be a unital commutative k-algebra such that

$A \otimes A \to A$

$a \otimes b \to ab$

is an isomorphism of k-algebras.

Two typical (and frequent) cases:

(i) $A = k/I$, I an ideal of k;

(ii) $A = S^{-1}k$, a localization of k.

Identifying A^{n+1} with A via $(a_o,\dots,a_n) \to a_oa_1 \dots a_n$, our double chain complex $C(A)$ looks like this:

$$\begin{array}{lccccccc}
C(A)_{3,*} & A & \xleftarrow{2} & A & \xleftarrow{0} & A & \xleftarrow{2} & A \longleftarrow \\
 & \downarrow 0 & & \downarrow -1 & & \downarrow 0 & & \downarrow -1 \\
C(A)_{2,*} & A & \xleftarrow{0} & A & \xleftarrow{3} & A & \xleftarrow{0} & A \longleftarrow \\
 & \downarrow 1 & & \downarrow 0 & & \downarrow 1 & & \downarrow 0 \\
C(A)_{1,*} & A & \xleftarrow{2} & A & \xleftarrow{0} & A & \xleftarrow{2} & A \longleftarrow \\
 & \downarrow 0 & & \downarrow -1 & & \downarrow 0 & & \downarrow -1 \\
C(A)_{o,*} & A & \xleftarrow{0} & A & \xleftarrow{1} & A & \xleftarrow{0} & A \longleftarrow
\end{array}$$

(Note that with our identification $A^{n+1} = A$, $n \geq 0$, we obtain $t = (-1)^n$, hence $D = 1 - (-1)^n$ and $N = \begin{cases} 0 & n \text{ odd} \\ (n+1)1 & n \text{ even} \end{cases}$ on $A^{n+1} = A$)

The double cochain complex $C^t(A)$: replace A by $A^* = \mathrm{Hom}_k(A,k)$ and reverse all arrows.
We obtain for Hochschild (co)homology:

$$H_n(A) = \begin{cases} A & n = 0 \\ 0 & n \geq 1 \end{cases}$$

$$H^n(A,A^*) = \begin{cases} A^* & n = 0 \\ 0 & n \geq 1 \end{cases}$$

which gives in cyclic (co)homology by the long exact (co)homology sequences (2.4.6, 2.4.7 and 2.6.13, 2.6.14):

$$HC_n(A) = \begin{cases} A & n \text{ even} \\ 0 & n \text{ odd} \end{cases}$$

$$HC^n(A) = \begin{cases} A^* & n \text{ even} \\ 0 & n \text{ odd} \end{cases}$$

Note that the eventual non-acyclic behaviour of the rows (2-torsion, (n+1)-torsion) does not affect the result.
Let us generalize a little bit.
Let A be as before, and consider $M_r(A)$, the <u>k-algebra</u> of $r \times r$-matrices with coefficients in A.
Then ([C.E., p.172]):

$$H_n(M_r(A)) = H_n(M_r(k), M_r(A)), \quad n \geq 0$$

$$H^n(M_r(A), M_r(A)^*) = H^n(M_r(k), M_r(A)^*), \quad n \geq 0$$

But $M_r(k)$ is a separable k-algebra ([C.E., p.179]), hence

$H_n(M_r(k),-) = H^n(M_r(k),-) = 0$ for $n \geq 1$. This yields:

$$H_n(M_r(A)) = \begin{cases} A & n = 0 \\ 0 & n \geq 1 \end{cases}$$

$$H^n(M_r(A), M_r(A)^*) = \begin{cases} A^* & n = 0 \\ 0 & n \geq 1 \end{cases}$$

By the same long exact sequence argument as before we thus obtain:

$$HC_n(M_r(A)) = \begin{cases} A & n \text{ even} \\ 0 & n \text{ odd} \end{cases}$$

$$HC^n(M_r(A)) = \begin{cases} A^* = \mathrm{Hom}_k(A,k) & n \text{ even} \\ 0 & n \text{ odd} \end{cases}$$

For $A = k$ we have recovered and dualized 2.4.8. Consider now $k = \mathbb{Z}$, $A = \mathbb{Q}$. We get

$$HC_n(M_r(\mathbb{Q})) = \begin{cases} \mathbb{Q} & n \text{ even} \\ 0 & n \text{ odd} \end{cases}$$

whereas $HC^n(M_r(\mathbb{Q})) = 0$ for all $n \geq 0$

(Caution: $M_r(\mathbb{Q})$ is considered as a $\mathbb{Z}$-algebra: $\mathrm{Hom}_{\mathbb{Z}}(\mathbb{Q},\mathbb{Z}) = 0$!)

Analogously: $k = \mathbb{Z}$, $A = \mathbb{Z}/p$

$$HC_n(M_r(\mathbb{Z}/p)) = \begin{cases} \mathbb{Z}/p & n \text{ even} \\ 0 & n \text{ odd} \end{cases}$$

$$HC^n(M_r(\mathbb{Z}/p)) = 0 \quad \text{for all } n \geq 0$$

(once again: $\mathrm{Hom}_{\mathbb{Z}}(\mathbb{Z}/p, \mathbb{Z}) = 0$)

It is clear that this type of result holds for <u>any</u> integral domain k (which is not a field) together with $A = \mathrm{Quot}(k)$ or $A = k/I$, I a proper ideal of k. Thus (at least in our special setting) cyclic homology only depends on A, whereas cyclic cohomology heavily depends on the structure homomorphism $k \to A$.

<u>Proposition 2.6.16</u> Let A_1, A_2 be two unital associative k-projective k-algebras. Consider $A = A_1 \times A_1$, their direct product, and the projections $\pi_1: A \to A_1$, $\pi_2: A \to A_2$. Then

$$HC^*(A_1) \oplus HC^*(A_2) \xrightarrow[HC^*(\pi_2)]{HC^*(\pi_1)} HC^*(A)$$

is an isomorphism.

<u>Proof</u>. Dualize 2.4.10 (see [C.E., p.173] for the result in Hochschild cohomology).

Remark 2.6.17 The suspension operator $S: HC^n(A) \to HC^{n+2}(A)$ in explicit form.

We shall assume that $\mathbb{Q} \subset k$.

Recall 2.6.7: $C_\lambda^*(A)$ is the subcomplex of the Hochschild cochain complex $(\mathrm{Hom}_k(A^{*+1},k),b)$ defined by

$$C_\lambda^n(A) = \mathrm{Ker}(\mathrm{Hom}_k(A^{n+1},k) \xrightarrow{1-\iota} \mathrm{Hom}_k(A^{n+1},k)), \quad n \geq 0$$

(i.e. $\tau \in C_\lambda^n(A) \Leftrightarrow \tau(a_o,..,a_n) = (-1)^n \tau(a_n,a_o,..,a_{n-1})$ for all $(a_o,..,a_n) \in A^{n+1}$)

We have (by 2.6.8 and 2.6.12) a quasi-isomorphism $C_\lambda^*(A) \to {}^BC(A)$, which is given by the inclusions

$$C_\lambda^n(A) \hookrightarrow \mathrm{Hom}_k(A^{n+1},k) \hookrightarrow {}^BC(A)^n$$

Now, $S: HC^n(A) \to HC^{n+2}(A)$ is induced by the inclusion ${}^BC(A)^n \hookrightarrow {}^BC(A)^{n+2}$.

We want to define $S_\lambda: Z_\lambda^n(A) \to Z_\lambda^{n+2}(A)$ such that

$$\begin{array}{ccc} Z_\lambda^{n+2}(A) & \longrightarrow & {}^BC(A)^{n+2} \\ \uparrow S_\lambda & & \uparrow S \\ Z_\lambda^n(A) & \longrightarrow & {}^BC(A)^n \end{array}$$

commutes modulo coboundaries (on the right side).

$(Z_\lambda^n(A) = Z^n(A,A^*) \cap \mathrm{Ker}\, D = \mathrm{Ker}\, b \cap \mathrm{Ker}\, D = \left\{\begin{matrix}\text{cyclic} \\ \text{n-cocycles}\end{matrix}\right\})$

For $\tau \in Z_\lambda^n(A)$ define $\tau_+ \in \mathrm{Hom}_k(A^{n+2},k)$ by

$$\tau_+(a_o,\ldots,a_{n+1}) = \sum_{i=0}^{n} \sum_{K=0}^{i} (-1)^K \tau(a_o,\ldots,a_K a_{K+1},\ldots,a_{n+1})$$

and $S_\lambda \tau = -\frac{1}{(n+1)(n+2)} b\tau_+$

I claim that

(i) $S_\lambda \tau \in Z_\lambda^{n+2}(A)$ for $\tau \in Z_\lambda^n(A)$

(ii) $S_\lambda \tau - \tau = (b+B)(-\frac{1}{(n+1)(n+2)} \tau_+)$

(which will prove our assertion).

First step. $sD\tau_+ = (n+2)\tau$ for $\tau \in Z_\lambda^n(A)$

$$(sD\tau_+)(a_o,\dots,a_n) = (D\tau_+)(1,a_o,\dots,a_n)$$
$$= \tau_+(1,a_o,\dots,a_n) + (-1)^n\tau_+(a_n,1,a_o,\dots,a_{n-1})$$

But

$$\tau_+(1,a_o,..,a_n) + (-1)^n\tau_+(a_n,1,a_o,..,a_{n-1})$$
$$= \sum_{i=0}^{n} [(\tau(a_o,..,a_n)-\tau(1,a_oa_1,..,a_n)+\dots+(-1)^i\tau(1,a_o,..,a_{i-1}a_i,..,a_n))$$
$$+ (-1)^n(\tau(a_n,a_o,..,a_{n-1})-\tau(a_n,a_o,..,a_{n-1})+\tau(a_n,1,a_oa_1,..,a_{n-1})+\dots$$
$$+ (-1)^i\tau(a_n,1,a_o,..,a_{i-2}a_{i-1},..,a_n))]$$
$$= 2\tau(a_o,..,a_n) + \sum_{i=1}^{n} (\tau(a_o,..,a_n)+(-1)^i\tau(1,a_o,..,a_{i-1}a_i,..,a_n))$$
$$= (n+2)\tau(a_o,..,a_n)+(b\tau)(1,a_o,..,a_n)-\tau(a_o,..,a_n)-(-1)^{n+1}\tau(a_n,a_o,..,a_{n-1})$$
$$= (n+2)\tau(a_o,..,a_n) \quad (\text{since } \tau \in \operatorname{Ker} b \cap \operatorname{Ker} D).$$

Second step. $B\tau_+ = (n+1)(n+2)\tau$

Now, $B\tau_+ = NsD\tau_+$, i.e.

$$(B\tau_+)(a_o,..,a_n) = \sum_{K=0}^{n} (-1)^{nK}(sD\tau_+)(a_{n-K+1},\dots,a_n,a_o,..,a_{n-K})$$
$$= \sum_{K=0}^{n} (-1)^{nK}(n+2)\tau(a_{n-K+1},\dots,a_n,a_o,..,a_{n-K})$$
$$= (n+1)(n+2)\tau(a_o,..,a_n)$$

(since $\tau \in \operatorname{Ker} D$).

Finally: $S_\lambda\tau - \tau = (b+B)(-\frac{1}{(n+1)(n+2)}\tau_+)$ for $\tau \in Z_\lambda^n(A)$ which proves (ii). (i) is a consequence of (ii).

Remark 2.6.18 Recall the definition of the double (chain) complex $\mathcal{B}(A)$ (2.4.13). We can dualize and obtain thus the double cochain complex $\mathcal{B}^t(A)$, such that

$$(\operatorname{Tot}(\mathcal{B}^t(A)),d) = ({}^{B}C(A),d), \quad \text{i.e.}$$

$HC^n(A) = H^n(\mathrm{Tot}(\mathcal{B}^t(A)),d), \quad n \geq 0.$

Theorem 2.6.19 Let $(E_r)_{r\geq 1}$ be the spectral sequence associated with the second filtration of $T = \mathrm{Tot}(\mathcal{B}^t(A))$. Then

$$E_2^{p,q} \underset{p}{\Longrightarrow} HC^n(A), \quad (n = p+q)$$

and the following holds:

(1) $E_1^{p,q} = H^{q-p}(A,A^*), \quad q \geq p$

(2) $d_1^{p,q}: H^{q-p}(A,A^*) \to H^{q-p-1}(A,A^*)$

is induced by B.

Proof. The proof of 2.4.15 dualizes step by step.

Remark 2.6.20 Let A be a unital associative k-algebra which is k-projective. Then the projection

$$(A^{*+1},b) \to (A \otimes \bar{A}^*,b)$$

dualizes to an injection

$$(\mathrm{Hom}_k(A \otimes \bar{A}^*,k),b) \to (\mathrm{Hom}_k(A^{*+1},k)b)$$

which is a quasi-isomorphism.

This is seen by the following argument ([E.C., pp.174/176]): Write the acyclic Hochschild complex (A^{*+2},b') in the form $(A^e \otimes A^*,b')$. We get an A^e-projective resolution of A (since A is supposed to be k-projective). Now, when passing to $(A^e \otimes \bar{A}^*,b')$ (which makes sense), acyclicity still is valid (since the contracting homotopy s passes to the quotient), and we get thus another A^e-projective resolution of A. We have

$$(\mathrm{Hom}_k(A \otimes \bar{A}^*,k),b) = (\mathrm{Hom}_{A^e}(A^e \otimes \bar{A}^*,\mathrm{Hom}_k(A,k)),\mathrm{Hom}(b',1))$$

$$(\mathrm{Hom}_k(A^{*+1},k),b) = (\mathrm{Hom}_{A^e}(A^e \otimes A^*,\mathrm{Hom}_k(A,k)),\mathrm{Hom}(b',1))$$

and hence our injection is a quasi-isomorphism by the standard homotopy equivalence argument for projective resolutions.

Consider the normalized mixed Hochschild cochain complex

$(\overline{C}(A),b,B) = (\mathrm{Hom}_k(A \otimes \overline{A}^*,k),b,B)$

where the operators b and B are merely restrictions of b and B on $C(A) = \mathrm{Hom}_k(A^{*+1},k)$ (or, equivalently, induced by dualizing b and B on $\overline{C}(A) = (A \otimes \overline{A}^*,b,B)$).

Proposition 2.6.21 $HC^n(A) = H^n({}^{B}\overline{C}(A),d)$, $n \geq 0$.

Proof. 2.6.10(6) together with 2.6.20 (compare with the argument in the proof of 2.5.3).

Remark 2.6.22 We shall not define nor discuss reduced cyclic cohomology. The machinery is obviously available since the dualizing arguments to be applied on 2.5.6 ... (until 2.5.18) should be clear.

I.2.7 Morita-invariance of Hochschild homology and of cyclic homology.

Example 2.7.1 Let k be an arbitrary commutative ring, and let A be a unital associative k-algebra, $B = M_r(A)$, $r \geq 1$, the k-algebra of $r\times r$-matrices with coefficients in A.
Consider $P = {}^tA^r$, the left A-module of $1\times r$-matrices (rows) with coefficients in A, and $Q = A^r$, the left B-module of $r\times 1$-matrices (columns) with coefficients in A. P is a right B-module, and Q is a right A-module, in a natural fashion.
Moreover, the left and right actions on P and on Q are associatively compatible; hence P is an A-B bimodule (equivalently: a left $A\otimes B^o$-module; $(\ )^o$ denotes opposite multiplication), and Q is a B-A bimodule (a left $B\otimes A^o$-module). Note that $P \otimes_B Q \simeq A$ as an A-A bimodule via scalar product multiplication of rows with columns, and $Q \otimes_A P \simeq B$ as a B-B bimodule via Kronecker product multiplication of columns with rows (identifying $e_i \otimes {}^te_j$ with e_{ij}, $1 \leq i, j \leq r$, for the usual standard A-bases $(e_1,\dots,e_r)$ of Q, $({}^te_1,\dots,{}^te_r)$ of P and $(e_{11},\dots,e_{rr})$ of $B = M_r(A)$).
Furthermore: P is A-projective (since it is A-free) as well as B-projective (since it is a B-direct summand in B), but not necessarily $A\otimes B^o$-projective. Similarly: Q is projective over both rings, but not necessarily $B\otimes A^o$-projective.

Definition 2.7.2 Let A and B be two unital associative rings (unital associative k-algebras for some commutative ring k). A and B are said to be <u>Morita-equivalent</u> if there is an A-B bimodule P and a B-A bimodule Q such that $P \otimes_B Q \simeq A$ as an A-A bimodule, and $Q \otimes_A P \simeq B$ as a B-B-bimodule.

<u>Remark 2.7.3</u> Let P be a left $A \otimes B^o$-module.
The following conditions are equivalent:

(a) $(-) \otimes_A P$: Mod-A → Mod-B
is an equivalence of categories (between right A-moduled and right B-modules).

(b) There is a left $B \otimes A^o$-module Q such that $P \otimes_B Q \simeq A$ and $Q \otimes_A P \simeq B$ as bimodules.

(c) $P \otimes_B (-)$: B-Mod → A-Mod is an equivalence of categories.

(cf. [Ba, p.60]).

<u>Complement 2.7.4</u>

(1) Using the identification $(P \otimes_B Q) \otimes_A P = P \otimes_B (Q \otimes_A P)$ and writing pq for the image of $p \otimes q$ in A, qp for the image of $q \otimes p$ in B, we may assume that $(pq)p' = p(qp')$ for the left and right actions on P; and similarly for Q.

(2) Let now P and Q be as in the definition 2.7.2. Then P and Q are necessarily finitely generated and projective as A-modules as well as B-modules. Let us show this for P as a B-module: Write $1 = \sum_{i=1}^{N} p_i q_i$ in A (with the "scalar product" meaning of pq as indicated in (1)).

Consider $\alpha: P \to B^N$, given by
$$p \to (q_1 p, \ldots, q_N p)$$
α is a B-homomorphism.

Let $\beta: B^N \to P$ be defined by
$$\beta(b_1, \ldots, b_N) = \sum_{i=1}^{N} p_i b_i$$
(note that $p_1, \ldots, p_N, q_1, \ldots, q_N$ are given by the fixed partition of the unity in A above).
Then $\beta\alpha = id_P$ by the associativity property of (1), hence P is

finitely generated and projective as a B-module.

Definition 2.7.5 Let k be a commutative ring, A a unital associative k-algebra, and M and A-A bimodule. For $n \geq 0$, set

$$C_n(A,M) = M \otimes A^n = M \otimes A \otimes A \otimes \ldots \otimes A \quad (n \text{ copies of } A)$$

$b: C_n(A,M) \to C_{n-1}(A,M)$ is given by the formula

$$\begin{aligned} b(m \otimes (a_1,\ldots,a_n)) &= ma_1 \otimes (a_2,\ldots,a_n) \\ &+ \sum_{i=1}^{n-1} (-1)^i m \otimes (a_1,\ldots,a_i a_{i+1},\ldots,a_n) \\ &+ (-1)^n a_n m \otimes (a_1,\ldots,a_{n-1}). \end{aligned}$$

The chain complex $(C_*(A,M),b)$ is called the Hochschild complex of A with coefficients in M. $H_n(A,M)$, its n-th homology group (which is a k-module), $n \geq 0$, is called the n-th Hochschild homology of A with coefficients in M.

Remark 2.7.6

(1) For $A = M$ we have $(C_*(A,A),b) = (A^{*+1},b)$, our usual Hochschild complex of A.

(2) Assume A to be k-flat.
Then $H_*(A,M) = \mathrm{Tor}_*^{A^e}(A,M)$, by the same argument as in 2.1.2(2).

Lemma 2.7.7 In the situation 2.7.5, let M be a left A-module, Ω a projective right A-module.

Then $H_n(A,M \otimes \Omega) = \begin{cases} \Omega \otimes_A M & \text{for } n = 0 \\ 0 & \text{for } n \geq 1 \end{cases}$

Proof.

(i) We shall first treat the particular case $\Omega = A$. We have to show:

$$H_n(A,M \otimes A) = \begin{cases} M & n = 0 \\ 0 & n \geq 1 \end{cases}$$

Consider the augmentation map

$$\varepsilon: C_0(A,M \otimes A) = M \otimes A \to M$$

given by $\varepsilon(m \otimes a) = am$.

We obtain a chain contraction for the augmented complex when defining

$$\eta(m) = m \otimes 1$$

$$s((m \otimes a) \otimes (a_1,\ldots,a_n)) = (m \otimes 1) \otimes (a,a_1,\ldots,a_n)$$

This yields the assertion.

(ii) The general case is easily reduced to (i), since $\Omega \underset{A}{\otimes} (-)$ is exact, and since

$$C_*(A,M \otimes \Omega) \simeq \Omega \underset{A}{\otimes} C_*(A,M \otimes A)$$

via the isomorphisms

$$(M \otimes \Omega) \otimes A^n \simeq \Omega \underset{A}{\otimes} (M \otimes A \otimes A^n)$$

where we need only make explicit the left A-module structure on $M \otimes A \otimes A^n$:

$$x.(m \otimes a) \otimes (a_1,\ldots,a_n) = (m \otimes xa) \otimes (a_1,\ldots,a_n)$$

<u>Complement 2.7.8</u> Let M be a right A-module, and let P be a projective left A-module. Then

$$H_n(A,P \otimes M) = \begin{cases} M \underset{A}{\otimes} P & \text{for } n = 0 \\ 0 & \text{for } n \geq 1 \end{cases}$$

<u>Theorem 2.7.9</u> Let k be any commutative ring, and let A and B be unital associative k-algebras, P an A-B bimodule which is projective over both rings, and Ω any B-A bimodule.

Then there is a natural sequence of isomorphisms

$$F_n:\ H_n(A,P \underset{B}{\otimes} \Omega) \to H_n(B,\Omega \underset{A}{\otimes} P), \quad n \geq 0$$

which vary functorially with the 4-tuple $(A,B;P,\Omega)$.

<u>Proof</u>. Consider the following double complex $(C_{p,q},d',d'')$:

$$C_{p,q} = P \otimes B^q \otimes \Omega \otimes A^p = C_p(A,P \otimes B^q \otimes \Omega) \simeq C_q(B,\Omega \otimes A^p \otimes P), \quad p,q \geq 0$$

where the last isomorphism is given by cyclic permutation of the relevant terms.

$d'_{p,q}: C_{p,q} \to C_{p-1,q}$ is the boundary map b for the Hochschild complex $C_*(A, P \otimes B^q \otimes Q)$, whereas $d''_{p,q}: C_{p,q} \to C_{p,q-1}$ is equal to $(-1)^p b$, the boundary map (up to a sign) for the Hochschild complex $C_*(B, Q \otimes A^p \otimes P)$.

The columns of C_{**} are Hochschild complexes for the homology of A with coefficients in certain A-A bimodules parametrized by B, and the rows of C_{**} are Hochschild complexes for the homology of B with coefficients in certain B-B bimodules parametrized by A.

Let us draw a picture:

$$\begin{array}{lccccc}
C_{3,*}: & P \otimes Q \otimes A^3 & \xleftarrow{-b} & P \otimes B \otimes Q \otimes A^3 & \xleftarrow{-b} & P \otimes B^2 \otimes Q \otimes A^3 \\
& \downarrow b & & \downarrow b & & \downarrow b \\
C_{2,*}: & P \otimes Q \otimes A^2 & \xleftarrow{b} & P \otimes B \otimes Q \otimes A^2 & \xleftarrow{b} & P \otimes B^2 \otimes Q \otimes A^2 \\
& \downarrow b & & \downarrow b & & \downarrow b \\
C_{1,*}: & P \otimes Q \otimes A & \xleftarrow{-b} & P \otimes B \otimes Q \otimes A & \xleftarrow{-b} & P \otimes B^2 \otimes Q \otimes A \\
& \downarrow b & & \downarrow b & & \downarrow b \\
C_{0,*}: & P \otimes Q & \xleftarrow{b} & P \otimes B \otimes Q & \xleftarrow{b} & P \otimes B^2 \otimes Q \\
& & & & & \\
& C_{*,0} & & C_{*,1} & & C_{*,2}
\end{array}$$

It is immediate that $d'^2 = d''^2 = 0$.

The property $d'd'' + d''d' = 0$ follows easily:

Consider $p \otimes (b_1,\dots,b_m) \otimes q \otimes (a_1,\dots,a_n) \in P \otimes B^m \otimes Q \otimes A^n$.

We get: $(-1)^{n-1} d''d'(p \otimes (b_1,\dots,b_m) \otimes q \otimes (a_1,\dots,a_n))$

$$\begin{aligned}
&= (-1)^n d'd''(p \otimes (b_1,\dots,b_m) \otimes q \otimes (a_1,\dots,a_n)) \\
&= pb_1 \otimes (b_2,\dots,b_m) \otimes qa_1 \otimes (a_2,\dots,a_n) \\
&+ (-1)^n a_n p b_1 \otimes (b_2,\dots,b_m) \otimes q \otimes (a_1,\dots,a_{n-1}) \\
&+ (-1)^m p \otimes (b_1,\dots,b_{m-1}) \otimes b_m q a_1 \otimes (a_2,\dots,a_n) \\
&+ (-1)^{m+n} a_n p \otimes (b_1,\dots,b_{m-1}) \otimes b_m q \otimes (a_1,\dots,a_{n-1}) \\
&+ \sum_{i=1}^{n-1} (-1)^i [pb_1 \otimes (b_2,\dots,b_m) \otimes q \otimes (a_1,\dots,a_i a_{i+1},\dots,a_n) \\
&\qquad + (-1)^m p \otimes (b_1,\dots,b_{m-1}) \otimes b_m q \otimes (a_1,\dots,a_i a_{i+1},\dots,a_n)]
\end{aligned}$$

$$+\sum_{j=1}^{m-1}(-1)^j[p\otimes(b_1,\ldots,b_jb_{j+1},\ldots,b_m)\otimes qa_1\otimes(a_2,\ldots,a_n)$$

$$+(-1)^n a_np\otimes(b_1,\ldots,b_jb_{j+1},\ldots,b_m)\otimes q\otimes(a_1,\ldots,a_{n-1})]$$

$$+\sum_{i=1}^{n-1}\sum_{j=1}^{m-1}(-1)^{i+j}p\otimes(b_1,\ldots,b_jb_{j+1},\ldots,b_m)\otimes q\otimes(a_1,\ldots,a_ia_{i+1},\ldots,a_n)$$

Let us look now at the two standard filtrations of T, the total complex associated with (C_{**},d',d'').

Recall: $T_n = \bigoplus_{p+q=n} C_{p,q}$, $n \geq 0$

and $d_n: T_n \to T_{n-1}$ is given by $d_n = \sum_{p+q=n}(d'_{p,q}+d''_{p,q})$.

By virtue of 1.4.11 and 1.4.12 we have

$${}^{I}E^2_{p,q} = H'_pH''_{p,q}(C_{**}) \quad \text{and}$$

$${}^{II}E^2_{p,q} = H''_pH'_{q,p}(C_{**}).$$

Our assumptions on the two-sided projectivity of P, together with 2.7.7 and 2.7.8, yield:

$$H''_{p,q}(C_{**}) = H_q(B,Q\otimes A^p\otimes P) = \begin{cases} P\underset{B}{\otimes}Q\otimes A^p & \text{for } q=0 \\ 0 & \text{for } q\geq 1\end{cases}$$

$$H'_{q,p}(C_{**}) = H_q(A,P\otimes B^p\otimes Q) = \begin{cases} B^p\otimes Q\underset{A}{\otimes}P & \text{for } q=0 \\ 0 & \text{for } q\geq 1\end{cases}$$

Consequently we obtain:

$${}^{I}E^2_{p,q} = \begin{cases} H_p(A,P\underset{B}{\otimes}Q) & \text{for } q=0 \\ 0 & \text{for } q\geq 1\end{cases}$$

$${}^{II}E^2_{p,q} = \begin{cases} H_p(B,Q\underset{A}{\otimes}P) & \text{for } q=0 \\ 0 & \text{for } q\geq 1\end{cases}$$

Now, by 1.4.13, we get simultaneously

$${}^{I}E^2_{n,o} \simeq H_n(T) \simeq {}^{II}E^2_{n,o}$$

i.e. $H_n(A,P\underset{B}{\otimes}Q) \simeq H_n(B,Q\underset{A}{\otimes}P)$ for all $n \geq 0$.

Furthermore (look at the second part of the proof of 2.2.6) the isomorphisms are induced by "projection from the diagonal on the outer factors".
It is now clear that our isomorphisms are functorial in $(A,B;P,Q)$, since the double comple C_{**} is functorial in these 4 variables.

Corollary 2.7.10 Let A and B be two unital associative k-algebras which are Morita-equivalent. Then there is a natural sequence of isomorphisms in Hochschild homology:

F_n: $H_n(A) \to H_n(B)$, $n \geq 0$

which are functorial in A and B.

Remark 2.7.11 k any commutative ring, A a unital associative k-algebra, $B = M_r(A)$, $r \geq 1$, the k-algebra of $r \times r$-matrices with coefficients in A.
Note that $M_r(A) \simeq M_r(k) \otimes A$
via the identification $(a_{ij}) = \Sigma e_{ij} \otimes a_{ij}$
(where $\{e_{ij}: 1 \leq i, j \leq r\}$ is the standard k-basis of $M_r(k)$).

Define $Tr: C(M_r(A) \to C(A)$

by the formula:

$$Tr((x_o \otimes a_o),\ldots,(x_n \otimes a_n)) = tr(x_o \cdot \ldots \cdot x_n)(a_o,\ldots,a_n)$$

where $x_i \in M_r(k)$, $a_i \in A$, $0 \leq i \leq n$

(and where $tr: M_r(k) \to k$ is the usual trace-form).

<u>Claim</u>. $Tr: C(M_r(A)) \to C(A)$ is a homomorphism of mixed complexes

(1) $Tr \circ b = b \circ Tr$:

It is immediate that

$$(Tr \circ b)((x_o \otimes a_o),\ldots,(x_n \otimes a_n)) = (b \circ Tr)((x_o \otimes a_o),\ldots,(x_n \otimes a_n))$$
$$= tr(x_o \cdot \ldots \cdot x_n) b(a_o,\ldots,a_n)$$

(2) $Tr \circ B = B \circ Tr$:

In complete analogy to (1) we have

$$(\mathrm{Tr}\circ B)((x_o \otimes a_o),\dots,(x_n \otimes a_n)) = (B\circ \mathrm{Tr})((x_o \otimes a_o),\dots,(x_n \otimes a_n))$$

$$= \mathrm{tr}(x_o\cdot\ldots\cdot x_n)B(a_o,\dots,a_n)$$

Complement 2.7.12 (traces and domino-indexing)

$\mathrm{Tr}_n: (M_r(A))^{n+1} \to A^{n+1}$ is given by the formula:

$$\mathrm{Tr}_n(m^o,m^1,\dots,m^n) = \sum_{(i_o,i_1,\dots,i_n)} (m^o_{i_o i_1}, m^1_{i_1 i_2},\dots,m^n_{i_n i_o})$$

where m^K_{ij} means the (i,j)-entry of the matrix m^K, and where the summation goes over all $(n+1)$-tuples $(i_o,i_1,\dots,i_n) \in \{1,\dots,r\}^{n+1}$.

Note the "domino-compatibility" of the subscripts in the sum.
We have only to make sure that our formula is correct for

$$(m^o,m^1,\dots,m^n) = (e^{a_o}_{i_o j_o}, e^{a_1}_{i_1 j_1},\dots,e^{a_n}_{i_n j_n})$$

where $e^a_{ij} = e_{ij} \otimes a$, $a \in A$, $1 \le i, j \le r$.
But this is clear, since

$$\mathrm{tr}(e_{i_o j_o}\cdot\ldots\cdot e_{i_n j_n})(a_o,\dots,a_n) = \begin{cases} (a_o,\dots,a_n) & \text{whenever } j_o = i_1, \\ & j_1 = i_2,\dots,j_n = i_o \\ 0 & \text{otherwise} \end{cases}$$

and, equally

$$\sum_{(i_o,\dots,i_n)} (m^o_{i_o i_1}, m^1_{i_1 i_2},\dots,m^n_{i_n i_o}) = \begin{cases} (a_o,\dots,a_n) & \text{whenever } j_o = i_1, \\ & j_1 = i_2,\dots,j_n = i_o \\ 0 & \text{otherwise} \end{cases}$$

Note that in particular

$\mathrm{Tr}_o: M_r(A) \to A$ is given by

$$\mathrm{Tr}_o((a_{ij})) = \sum_{i_o=1}^{r} a_{i_o i_o},$$ as expected.

Proposition 2.7.13 Let A be a unital associative k-algebra, M an A-A bimodule.

$$\mathrm{Tr}_*: C_*(M_r(A),M_r(M)) \to C_*(A,M)$$

given by

$$Tr_n(m,x^1,\dots,x^n) = \sum_{(i_o,\dots,i_n)} (m_{i_o i_1}, x^1_{i_1 i_2}, \dots, x^n_{i_n i_o}), \quad n \geq 0$$

is a quasi-isomorphism (i.e. induces an isomorphism in Hochschild homology).

Proof. We shall place ourselves in the situation 2.7.9:
First, consider the 4-tuple $(A,B;P,Q) = (A,A;A,M)$, then the 4-tuple $(A,B;P,Q) = (M_r(A),A;A^r,M^r)$, where A^r is the $M_r(A)$-A bimodule of all $r\times 1$-matrices (columns) with coefficients in A, whereas M^r is the A-$M_r(A)$ bimodule of all $1\times r$ matrices (rows) with coefficients in M.
We have to exploit the functoriality of the isomorphisms

$$F_n: H_n(A,P \otimes_B Q) \to H_n(B,Q \otimes_A P)$$

relative to the set of variables $(A,B;P,Q)$.

Define $G: (A,A;A,M) \to (M_r(A),A;A^r,M^r)$

by $\quad G(a,b;p,m) = \left(D(a),b;\begin{pmatrix}p\\o\\\vdots\\o\end{pmatrix},(m,0,\dots,0)\right)$

where

$$D(a) = \begin{pmatrix} a & & 0\\ & a & \\ & & \ddots \\ 0 & & a\end{pmatrix} = a\mathrm{Id}$$

is the diagonal matrix defined by $a \in A$.
It is immediate that the first two components are defined by homomorphisms of k-algebras, and that the second two components are given by homomorphisms of bimodules (where the A-A structures on A and M are induced by the $M_r(A)$-A structure on A^r and by the A-$M_r(A)$ structure on M^r via pullback along $D: A \to M_r(A)$).
Thus we get a commuting square

$$\begin{array}{ccc} H_n(A,M) & \xrightarrow[\sim]{F_n} & H_n(A,M) \\ \downarrow G_1 & & \downarrow G_2 \\ H_n(M_r(A),M_r(M)) & \xrightarrow[\sim]{F'_n} & H_n(A,M) \end{array}$$

since, for the lower row, we have:

$$A^r \otimes_A M^r = M_r(M) \quad \text{(Kronecker product of columns with rows)}$$

$$M^r \underset{M_r(A)}{\otimes} A^r = M \quad \text{(scalar product of rows with columns)}$$

Now, look first at G_2.

In the ring-variable, G_2 is given by the identity $A \to A$.

In the bimodule-varialbe, G_2 is induced by the following composition of maps

$$\begin{array}{ccccccc} A & \longrightarrow & M \underset{A}{\otimes} A & \longrightarrow & M^r \underset{M_r(A)}{\otimes} A^r & \longrightarrow & M \\ m & \longrightarrow & m \otimes 1 & \longrightarrow & (m,0,..,0) \otimes \begin{pmatrix} 1 \\ 0 \\ \vdots \\ 0 \end{pmatrix} & \longrightarrow & m \end{array}$$

hence by the identity, too.

Thus G_2 is the identity map, and consequently G_1 is an isomorphism.

Let us look now more closely at G_1.

In the ring-variable, G_1 is given by

$$D: A \longrightarrow A_r(A)$$

$$a \to \begin{pmatrix} a & & 0 \\ & a & \\ & & \ddots \\ 0 & & a \end{pmatrix}$$

In the bimodule-variable, G_1 is induced by

$$\begin{array}{ccccccc} M & \longrightarrow & A \underset{A}{\otimes} M & \longrightarrow & A^r \underset{A}{\otimes} M^r & \longrightarrow & M_r(M) \\ m & \longrightarrow & 1 \otimes m & \longrightarrow & \begin{pmatrix} 1 \\ 0 \\ \vdots \\ 0 \end{pmatrix} \otimes (m,0,\ldots,0) & \longrightarrow & \begin{pmatrix} m & 0 & \ldots & 0 \\ 0 & & & \\ \vdots & & 0 & \\ 0 & & & \end{pmatrix} = T(m) \end{array}$$

This gives the explicit formula for G_1 on the level of Hochschild complexes:

$$g_n: \ C_n(A,M) \longrightarrow C_n(M_r(A),M_r(M))$$

$$g_n(m \otimes (a_1,\ldots,a_n)) = T(m) \otimes (D(a_1),\ldots,D(a_n))$$

It is immediate that $Tr_n \circ g_n = id_{C_n(A,M)}$, $n \geq 0$.

But the homomorphism of chain complexes $(g_n)_{n\geq 0}$ is a quasi-isomorphism (since G_1 is an isomorphism). Hence

$$Tr_*: C_*(M_r(A), M_r(M)) \to C_*(A,M)$$

is a quasi-isomorphism, too.
This finishes the proof of our proposition.

Corollary 2.7.14 (Morita-invariance of cyclic homology)
The homomorphism of mixed complexes

$$Tr_*: C(M_r(A)) \to C(A)$$

is a quasi-isomorphism, both in Hochschild homology and in cyclic homology.

Proof. 2.7.13 combined with 2.3.15.

Complement 2.7.15 For all $r \geq 1$, and all $s \geq r$ the following diagram of homomorphisms of mixed complexes is commutative:

$$\begin{array}{lll} Tr_*: & C(M_s(A)) & \searrow \\ & \uparrow & \quad C(A) \\ tr_*: & C(M_r(A)) & \nearrow \end{array}$$

(the vertical arrow is induced by the standard inclusion $M_r(A) \subset M_s(A)$, $r \leq s$).
We want to pass to the direct limit $M(A) = \varinjlim M_r(A)$, in order to establish an isomorphism

$$Tr_*: HC_*(M(A)) \to HC_*(A)$$

Note that $\varinjlim C(M_r(A))$ exists as a mixed complex, but $C(M(A))$ does not exist (since $M(A)$ is not unital, hence we cannot define the operator B).
But recall 2.4.5. We obtain a commutative square

$$\begin{array}{ccc} {}_BC(M_r(A)) & \xrightarrow{Tr_*} & {}_BC(A) \\ \downarrow f & & \downarrow f \\ Tot(C(M_r(A))) & \xrightarrow{Tr_*} & Tot(C(A)) \end{array}$$

where all arrows are quasi-isomorphisms.
This commutative square is compatible with the standard inclusions $M_r(A) \hookrightarrow M_s(A)$, $r \leq s$;

hence $Tr_*: HC_*(M(A)) \to HC_*(A)$

makes sense in the context of non-unital cyclic homology (cf. 2.5.14), and is clearly an isomorphism (since cyclic homology commutes with direct limits).

Comments on Chapter I.

The opening section on spectral sequences follows roughly [Ro]. A special emphasis is given to the approximation theorem 1.3.7, which allows a unified argumentation in the latter sections.
The exposition on cyclic homology is an amalgam and an extended version of parts of [Ka] and of [L.Qu.]. The section on Morita-invariance follows closely [Ig].
In more detail: The double complex approach to cyclic homology (I.2.1) is due to B.L. Tsygan [Ts], the mixed complex formulation to D. Burghelea [Ka]. The equivalence of these two approaches with the down-to-earth definition of A. Connes [Co] (in characteristic zero; see 2.2.6 and 2.4.5) has been established by J.L. Loday and D. Quillen [L.Qu.]. Cyclic homology as a derived functor (2.3.10): this is a result of C. Kassel [Ka], dualizing and simplifying an idea of A. Connes. The section on non-unital and reduced cyclic homology, important for applications and computations, is entirely due to J.L. Loday and D. Quillen [L.Qu.].

References to chapter I.

[Ba] Bass , H.: Algebraic K-theory. New York, Amsterdam: Benjamin 1968

[C.E.] Cartan, H., Eilenberg, S.: Homological Algebra. Princeton, N.J.: Princeton Univ. Press 1956

[Co] Connes, A.: Non-Commutative Differential Geometry. I.H.E.S. Publ. Math. vol.62. 1985, 41-144

[Go] Godement, R.: Topologie algébrique et théorie des faisceaux. Paris: Hermann 1964

[H.St.] Hilton, P.J., Stammbach, U.: A Course in Homological Algebra. New York, Heidelberg, Berlin: Springer 1971

[Ig] Igusa, K.: What happens to Hatcher and Wagoner's formula for $\pi_o C(M)$ when the first Postnikov invariant of M is non-trivial? Preprint Brandeis Univ.

[Ka] Kassel, C.: Cyclic homology, comodules and mixed complexes. Preprint (1985)

[L.Qu.] Loday, J.L, Quillen, D.: Cyclic homology and the Lie algebra homology of matrices. Comment. Math. Helvetici 59 (1984), 565-591

[ML] MacLane, S.: Homology. Berlin-Göttingen-Heidelberg: Spinger 1963

[Ro] Rotman, J.J.: An Introduction to Homological Algebra. New York...: Academic Press 1979

[Ts] Tsygan, B.L.: Homology of matrix Lie algebras over rings and Hochschild homology. Uspekhi Math. Nauk vol.38 (1983), 217-218

Chapter II. Particularities in characteristic zero

II.1 Relation to de Rham theory.

II.1.1 A first approach: Noncommutative de Rham complexes

Remark-Definition 1.1.1 Let k be a commutative ring, A a unital associative k-algebra. Define

$$\Omega_0 A = A, \qquad \Omega_1 A = \mathrm{Ker}\begin{pmatrix} A \otimes A & \xrightarrow{b'} & A \\ a \otimes b & \to & ab \end{pmatrix}$$

$\Omega_0 A$ and $\Omega_1 A$ are naturally A-A-bimodules.

Consider $d^o: A \to \Omega_1 A$ defined by $d^o(a) = 1 \otimes a - a \otimes 1$; d^o is a k-derivation from A to $\Omega_1 A$, $a \in A$, i.e. we have $d^o(ab) = (d^o a)b + a d^o(b)$ for all $a, b \in A$.

Moreover, this k-derivation is universal in the following sense: For every A-A-bimodule M and for any k-derivation $d: A \to M$ there is a unique homomorphism of A-A-bimodules $f: \Omega_1 A \to M$ such that $d = f \circ d^o$, i.e. such that the following diagram is commutative:

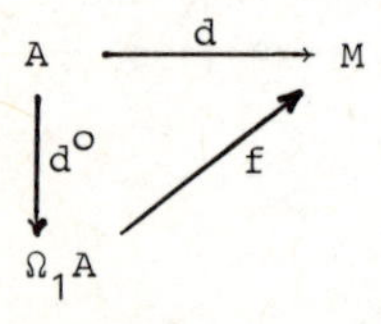

(cf. [C.E., p.168] for details)

For $n \geq 2$ define

$$\Omega_n A = \Omega_1 A \underset{A}{\otimes} \Omega_1 A \underset{A}{\otimes} \cdots \underset{A}{\otimes} \Omega_1 A \quad \text{(n-fold tensor product of } \Omega_1 A \text{ over } A\text{!).}$$

$\Omega_n A$ is an A-A-bimodule (operation on the exterior factors).

The elements of $\Omega_n A$ are called (noncommutative) k-differential forms

of degree n on A.
Note that $\Omega_1 A$ is generated by $d^o A$ as a left A-module (equally as a right A-module):
Take $\Sigma a_i \otimes b_i \in A \otimes A$ such that $\Sigma a_i b_i = 0$; then

$$\Sigma a_i \otimes b_i = \Sigma a_i (1 \otimes b_i - b_i \otimes 1) = \Sigma a_i d^o(b_i)$$

$$= \Sigma (a_i \otimes 1 - 1 \otimes a_i) b_i = -\Sigma d^o(a_i) b_i .$$

It follows that every element of $\Omega_n A$, $n \geq 1$, can be written as a finite sum of elements of the form

$$a_o d^o a_1 \otimes d^o a_2 \otimes \ldots \otimes d^o a_n \quad \text{with} \quad a_o, a_1, \ldots, a_n \in A$$

(note that in the definition of $\Omega_n A$ the n-fold tensor product is taken over A!).

Define $\quad \Omega_* A = \bigoplus_{n \geq 0} \Omega_n A.$

$\Omega_* A$ is a graded associative k-algebra (multiplication via tensoring), which is generated (as a k-algebra) by $A = \Omega_o A$ and $d^o A \subset \Omega_1 A$.

<u>Lemma 1.1.2</u> There is a unique graded k-derivation

$d\colon \Omega_* A \to \Omega_* A$ of degree +1, extending

$d^o\colon \Omega_o A \to \Omega_1 A$ such that $d^2 = 0$.

(Recall that we must have

$d(\omega_p \omega_n) = d\omega_p \omega_n + (-1)^p \omega_p d\omega_n$ for all $\omega_p \in \Omega_p A$, $\omega_n \in \Omega_n A$)

<u>Proof</u>. Uniqueness is easy: All elements of $\Omega_n A$ are k-linear combinations of elements of the form $a_o d^o a_1 \ldots\ldots d^o a_n$, $a_o, a_1, \ldots, a_n \in A$.
Now, since $d^2 = 0$, and since d extends $d^o\colon \Omega_o A \to \Omega_1 A$, we must have $d(a_o d^o a_1 \ldots d^o a_n) = d^o a_o d^o a_1 \ldots d^o a_n$, i.e. d is uniquely determined by d^o.
As to the existence of d, we first construct $d^1\colon \Omega_1 A \to \Omega_2 A$ such that $d^1 \circ d^o = 0$, and such that

$d^1(a\omega_1) = (d^o a)\omega_1 + a d^1 \omega_1$

$d^1(\omega_1 a) = (d^1 \omega_1) a - \omega_1 d^o a \quad$ for all $a \in A$, $\omega_1 \in \Omega_1 A$

Define $d^1 \colon A \otimes A \to \Omega_1 A \underset{A}{\otimes} \Omega_1 A = \Omega_2 A$

by the formula

$d^1(\Sigma a_i \otimes b_i) = \Sigma d^o a_i \otimes d^o b_i$

and restrict it to $\Omega_1 A \subset A \otimes A$.

$d^1 \colon \Omega_1 A \to \Omega_2 A$ is clearly k-linear.

$d^1 \circ d^o = 0$, since $d^1(d^o(a)) = d^1(1 \otimes a - a \otimes 1) = 0$ (note that $d^o(1)=0$)

Furthermore, consider $\omega_1 = \Sigma a_i \otimes b_i \in \Omega_1 A$, i.e. we have $\Sigma a_i \otimes b_i = 0$.

Then

$$\begin{aligned} d^1(a\omega_1) &= \Sigma d^o(aa_i) \otimes d^o b_i \\ &= \Sigma d^o(a)a_i \otimes d^o b_i + \Sigma a d^o(a_i) \otimes d^o b_i \\ &= d^o(a) \otimes \Sigma a_i d^o b_i + a d^1 \omega_1 \\ &= d^o(a)\omega_1 + a d^1 \omega_1 \end{aligned}$$

(note that $\Sigma a_i \otimes b_i = \Sigma a_i d^o b_i$)

Analogously

$$\begin{aligned} d^1(\omega_1 a) &= \Sigma d^o(a_i) \otimes d^o(b_i a) \\ &= \Sigma d^o(a_i) \otimes d^o(b_i)a + \Sigma d^o(a_i) \otimes b_i d^o(a) \\ &= (d^1\omega_1)a + (\Sigma d^o(a_i)b_i) \otimes d^o a \\ &= (d^1\omega_1)a - \omega_1 d^o a \end{aligned}$$

(recall that $\Sigma a_i \otimes b_i = -\Sigma d^o(a_i)b_i$)

This shows that $d^1 \colon \Omega_1 A \to \Omega_2 A$ has the required properties.

<u>Claim</u>. For all $n \geq 1$ there exists $d^n \colon \Omega_n A \to \Omega_{n+1} A$ such that

(1) $d^n \circ d^{n-1} = 0$

(2) $d^n(\omega_p \cdot \omega_q) = (d^p \omega_p) \cdot \omega_q + (-1)^p \omega_p \cdot d^q \omega_q$ for $\omega_p \in \Omega_p A$, $\omega_q \in \Omega_q A$,

$p + q = n$.

For $n = 1$ we just have given the proof.

Assume now $n > 1$, and that d^m has been constructed for $m < n$ (having properties (1) and (2)). Define

$\bar{d}^n: \Omega_1 A \times \Omega_{n-1}A \to \Omega_{n+1}A = \Omega_1 A \underset{A}{\otimes} \Omega_n A$ by

$$\bar{d}^n(\omega_1, \omega_{n-1}) = d^1\omega_1 \otimes \omega_{n-1} - \omega_1 \otimes d^{n-1}\omega_{n-1}$$

(note that the two $\otimes$-signs have different position in $\Omega_{n+1}A$)

$\bar{d}^n$ defines $d^n: \Omega_n A = \Omega_1 A \underset{A}{\otimes} \Omega_{n-1}A \to \Omega_{n+1}A$ provided that $\bar{d}^n(\omega_1 a, \omega_{n-1}) = \bar{d}^n(\omega_1, a\omega_{n-1})$ for all $a \in A$, $\omega_1 \in \Omega_1 A$, $\omega_{n-1} \in \Omega_{n-1}A$.

Now

$$\begin{aligned}\bar{d}^n(\omega_1 a, \omega_{n-1}) &= d^1(\omega_1 a) \otimes \omega_{n-1} - (\omega_1 a) \otimes d^{n-1}\omega_{n-1} \\ &= d^1\omega_1 \otimes a\omega_{n-1} - \omega_1 \otimes d^o a \otimes \omega_{n-1} - \omega_1 \otimes a d^{n-1}\omega_{n-1} \\ &= d^1\omega_1 \otimes a\omega_{n-1} - \omega_1 \otimes d^{n-1}(a\omega_{n-1}) \\ &= \bar{d}^n(\omega_1, a\omega_{n-1})\end{aligned}$$

Thus $d^n: \Omega_n A \to \Omega_{n+1}A$ is well-defined by

$$d^n(\omega_1 \otimes \omega_{n-1}) = d^1\omega_1 \otimes \omega_{n-1} - \omega_1 \otimes d^{n-1}\omega_{n-1}.$$

Let us show that $d^n \circ d^{n-1} = 0$.

Take $\omega_{n-1} = a d^o b \otimes \omega_{n-2} \in \Omega_{n-1}A$.

We have

$$\begin{aligned}d^{n-1}\omega_{n-1} &= d^1(a d^o b) \otimes \omega_{n-2} - a d^o b \otimes d^{n-2}\omega_{n-2} \\ &= d^o a \otimes d^o b \otimes \omega_{n-2} - a d^o b \otimes d^{n-2}\omega_{n-2}\end{aligned}$$

(since d^{n-1} satisfies (2), by the inductive hypothesis).

Thus

$$\begin{aligned}d^n d^{n-1}\omega_{n-1} &= -d^o a \otimes d^{n-1}(d^o b \otimes \omega_{n-2}) - d^1(a d^o b) \otimes d^{n-2}\omega_{n-2} \\ &\quad + a d^o b \otimes d^{n-1}d^{n-2}\omega_{n-2} \\ &= -d^o a \otimes (d^1 d^o b \otimes \omega_{n-2} - d^o b \otimes d^{n-2}\omega_{n-2}) - d^o a \otimes d^o b \otimes d^{n-2}\omega_{n-2} \\ &= 0\end{aligned}$$

Finally, (2) is satisfied for d^n in case $p = 1$, $q = n-1$ by definition, and follows in case $p = 0$, $q = n$ almost immediately. But for $p > 1$ $(q < n-1)$ (2) will follow by the inductive hypothesis:
Write $\omega_p \in \Omega_p A$, $p \geq 2$, in the form $\omega_p = \omega_1 \otimes \omega_{p-1} = \omega_1\omega_{p-1}$.
Then

$$\begin{aligned} d^n(\omega_p\omega_q) &= d^n(\omega_1\omega_{p-1}\omega_q) = d^1\omega_1 \otimes \omega_{p-1}\omega_q - \omega_1 \otimes d^{n-1}(\omega_{p-1}\omega_q) \\ &= (d^1\omega_1 \otimes \omega_{p-1}) \otimes \omega_q - \omega_1 \otimes (d^{p-1}\omega_{p-1}\cdot\omega_q + (-1)^{p-1}\omega_{p-1}d^q\omega_q) \\ &= (d^1\omega_1 \otimes \omega_{p-1} - \omega_1 \otimes d^{p-1}\omega_{p-1}) \otimes \omega_q + (-1)^p\omega_p d^q\omega_q \\ &= (d^p\omega_p)\omega_q + (-1)^p\omega_p d^q\omega_q, \quad \text{as claimed.} \end{aligned}$$

This finishes the proof of our lemma.

<u>Definition 1.1.3</u> k a commutative ring, A a unital associative k-algera.
$\Omega(A) = (\Omega_* A, d)$ is called the <u>differential envelope</u> of A.

<u>Example 1.1.4</u> The differential envelope $\Omega(A)$ of a tensor algebra $A = T(V) = \bigoplus_{m\geq 0} V^m$ (recall I.2.5.13).
We have already identified $\Omega_1 A = A \otimes V \otimes A$ (as an A-A-bimodule). Our universal k-derivation

$$d^o: A \to \Omega_1 A = A \otimes V \otimes A$$

reads like this (dot means tensoring inside A):
For a homogeneous element $v_1\cdot v_2\cdot\ldots\cdot v_m \in V^m \subset A$ we have

$$\begin{aligned} d^o(v_1\cdot v_2\cdot\ldots\cdot v_m) = 1 \otimes v_1 \otimes v_2\cdot v_3\cdot\ldots\cdot v_m + v_1 \otimes v_2 \otimes v_3\cdot\ldots\cdot v_m + \\ + v_1\cdot v_2 \otimes v_3 \otimes v_3 \otimes v_4\cdot\ldots\cdot v_m + \ldots \\ + v_1\cdot v_2\cdot\ldots\cdot v_{m-1} \otimes v_m \otimes 1. \end{aligned}$$

In particular: $d^o(v) = 1 \otimes v \otimes 1$ for $v \in V = V^1 \subset A$

We get

$\Omega_n A = A \otimes V \otimes A \otimes V \otimes \ldots \otimes V \otimes A$ (n times V and n+1 times A)

All tensor products are now over k, since we have already simplified ($A \otimes_A A = A$).

We have to make explicit $d\colon \Omega_* A \to \Omega_* A$, i.e.

we have to write out $d^n\colon \Omega_n A \to \Omega_{n+1} A$.

Following the construction in 1.1.2 we obtain first for

$$d^1\colon \Omega_1 A = A \otimes V \otimes A \quad \to \quad \Omega_2 A = A \otimes V \otimes A \otimes V \otimes A\colon$$

$$d^1(a \otimes v \otimes b) = d^o a \otimes v \otimes b - a \otimes v \otimes d^o b$$

(this is an immediate consequence of the definition given in 1.1.2; observe that $d^o a$ "visits" the first three factors of $\Omega_2 A$, whereas $d^o b$ "visits" the last three factors, of $\Omega_2 A$)

Now, this formula for d^1 easily generalizes:

$$d^n\colon \underset{(n \text{ times } V)}{A \otimes V \otimes A \otimes V \otimes \ldots \otimes V \otimes A} \to \underset{(n+1 \text{ times } V)}{A \otimes V \otimes A \otimes V \otimes \ldots \otimes V \otimes A}$$

is given by the following formula:

$$d^n(a_o \otimes v_1 \otimes a_1 \otimes \ldots \otimes v_n \otimes a_n) = \sum_{i=0}^{n} (-1)^i a_o \otimes v_1 \otimes \ldots \otimes d^o a_i \otimes \ldots \otimes v_n \otimes a_n$$

(i.e. d^o "visits" the n+1 A-arguments and "neglects" the n V-arguments) The formula is verified by induction on n (since d^n is defined inductively by the construction in 1.1.2).

<u>Proposition 1.1.5</u> $\Omega(A) = (\Omega_* A, d)$ has the following universal property: Let $Z = (\bigoplus_{j \geq 0} Z_j, \partial)$ be any d.g. (cochain) k-algebra (i.e. ∂ is a graded k-derivation of degree +1 such that $\partial^2 = 0$), and let $\varphi_o\colon A = \Omega_o A \to Z_o$ be a homomorphism of unital associative k-algebras. Then φ_o extends uniquely to a homomorphism $\varphi\colon \Omega(A) \to Z$ of d.g. (cochain) algebras.

<u>Proof</u>. Z_1 is an A-A-bimodule by pullback via φ_o, and $d = \partial \circ \varphi_o\colon A \to Z_1$ is a k-derivation from A to Z_1. Hence there exists a unique A-A-bimodule homomorphism $\varphi_1\colon \Omega_1 A \to Z_1$ such that the following square is commutative

$$\begin{array}{ccc} A = \Omega_o A & \xrightarrow{d^o} & \Omega_1 A \\ \downarrow \varphi_o & & \downarrow \varphi_1 \\ Z_o & \xrightarrow{\partial^o} & Z_1 \end{array}$$

We have explicitely:

$\varphi_1(a\omega_1) = \varphi_o(a)\varphi_1(\omega_1)$

$\varphi_1(\omega_1 a) = \varphi_1(\omega_1)\varphi_o(a)$ for all $a \in A$, $\omega_1 \in \Omega_1 A$

and $\varphi_1(d^o a) = \partial^o \varphi_o(a)$ for all $a \in A$.

We shall define (by induction on n) $\varphi_n : \Omega_n A \to Z_n$ such that

$\varphi_n(\omega_p \omega_q) = \varphi_p(\omega_p)\varphi_q(\omega_q)$, $\omega_p \in \Omega_p A$, $\omega_q \in \Omega_q A$, $p+q = n$

For $n = 1$ this has been done.

Assume $n > 1$, and suppose φ_m already defined for $m < n$ with the required multiplicative property.

Write $\Omega_n A = \Omega_1 A \otimes_A \Omega_{n-1} A$, and define

$\bar{\varphi}_n : \Omega_1 A \times \Omega_{n-1} A \to Z_n$

by $\bar{\varphi}_n(\omega_1, \omega_{n-1}) = \varphi_1(\omega_1)\varphi_{n-1}(\omega_{n-1})$.

We have $\bar{\varphi}_n(\omega_1 a, \omega_{n-1}) = \bar{\varphi}_n(\omega_1, a\omega_{n-1})$, hence we get

$\varphi_n : \Omega_n A = \Omega_1 A \otimes_A \Omega_{n-1} A \to Z_n$

$\varphi_n(\omega_1 \omega_{n-1}) = \varphi_1(\omega_1)\varphi_{n-1}(\omega_{n-1})$

As to the multiplicative property

$\varphi_n(\omega_p \omega_q) = \varphi_p(\omega_p)\varphi_q(\omega_q)$, $\omega_p \in \Omega_p A$, $\omega_q \in \Omega_q A$, $p+q = n$

it is immediate for $p = 0, 1$.

For $p \geq 2$ we decompose $\omega_p = \omega_1 \omega_{p-1}$, and we use the inductive hypothesis.

We thus get φ as a homomorphism of graded unital associative k-algebras. We have to make sure that φ respects the differentials, i.e. that for $n \geq 0$ the following square is commutative:

$$\begin{array}{ccc} \Omega_n A & \xrightarrow{d^n} & \Omega_{n+1} A \\ \downarrow \varphi_n & & \downarrow \varphi_{n+1} \\ Z_n & \xrightarrow{\partial^n} & Z_{n+1} \end{array}$$

Fot the first square $(n = 0)$ this has been seen to be true.
Let us look at the second square (i.e. $n = 1$).

Take $\omega_1 = ad^ob \in \Omega_1 A$, $a,b \in A$.

Now, $d^1(ad^ob) = (d^oa)d^ob$, and

$$_2(d^1(ad^ob)) = \varphi_2(d^oad^ob) = \varphi_1(d^oa)\varphi_1(d^ob)$$

y definition of φ_2.

n the other hand, $\varphi_1(ad^ob) = \varphi_o(a)\varphi_1(d^ob)$, hence

$$\begin{aligned} {}^1\varphi_1(ad^ob) &= \partial^1(\varphi_o(a)\varphi_1(d^ob)) \\ &= \partial^o\varphi_o(a)\varphi_1(d^ob) + \varphi_o(a)\partial^1\varphi_1(d^ob) \\ &= \varphi_1(d^oa)\varphi_1(d^ob). \end{aligned}$$

Since $\partial^1\varphi_1(d^ob) = \partial^1\partial^o\varphi_o(b) = 0$, and we are through.

Let us look at $n > 1$.

We can decompose $\omega_n = \omega_1\omega_{n-1} \in \Omega_n A = \Omega_1 A \underset{A}{\otimes} \Omega_{n-1}A$.

Then we get by the inductive hypothesis:

$$\begin{aligned} \varphi_{n+1}d^n\omega_n &= \varphi_{n+1}((d^1\omega_1)\omega_{n-1} - \omega_1 d^{n-1}\omega_{n-1}) \\ &= \varphi_2(d^1\omega_1)\varphi_{n-1}\omega_{n-1} - \varphi_1(\omega_1)\varphi_n(d^{n-1}\omega_{n-1}) \\ &= \partial^1(\varphi_1\omega_1)\varphi_{n-1}\omega_{n-1} - \varphi_1(\omega_1)\partial^{n-1}\varphi_{n-1}\omega_{n-1} \\ &= \partial^n((\varphi_1\omega_1)\varphi_{n-1}\omega_{n-1}) \\ &= \partial^n\varphi_n\omega_n \end{aligned}$$

The uniqueness of the extension of $\varphi_o: A \to Z_o$ to $\varphi: \Omega(A) \to Z$ follows from the uniqueness of $\varphi_1: \Omega_1 A \to Z_1$ together with the fact that A and $d^oA \subset \Omega_1 A$ generate $\Omega_* A$ as a k-algebra.

<u>Complement 1.1.6</u> Let $\varphi_o: A = \Omega_o A \to Z_o$ be surjective. Then its extension $\varphi: \Omega(A) \to Z$ is surjective if and only if Z_o and $\partial^o Z_o \subset Z_1$ generate Z as a k-algebra.

<u>Remark-Definition 1.1.7</u> Let $[\Omega_* A, \Omega_* A]$ be the graded k-submodule of

$\Omega_* A$ which is generated by all graded commutators, i.e. by all elements of the form $\omega_p\omega_q - (-1)^{pq}\omega_q\omega_p$, $\omega_p \in \Omega_p A$, $\omega_q \in \Omega_q A$, $p,q \geq 0$. We can write for the n-th homogeneous component:

$$[\Omega_* A,\Omega_* A]_n = \sum_{p+q=n} [\Omega_p A,\Omega_q A]$$

Define $\Lambda\Omega(A) = \Omega_* A/[\Omega_* A,\Omega_* A]$

$\Lambda\Omega(A)$ is graded via the quotient grading:

$$\Lambda^n\Omega(A) = \Omega_n A/\sum_{p+q=n} [\Omega_p A,\Omega_q A]$$

Furthermore, we have $d[\Omega_* A,\Omega_* A] \subset [\Omega_* A,\Omega_* A]$:

$$\begin{aligned} d(\omega_p\omega_q - (-1)^{pq}\omega_q\omega_p) &= (d\omega_p)\omega_q + (-1)^p\omega_p d\omega_q - (-1)^{pq}(d\omega_q)\omega_p \\ &\quad - (-1)^{pq}(-1)^q\omega_q d\omega_p \\ &= ((d\omega_p)\omega_q - (-1)^{(p+1)q}\omega_q d\omega_p) \\ &\quad + (-1)^p(\omega_p d\omega_q - (-1)^{p(q+1)}(d\omega_q)\omega_p) \end{aligned}$$

for $\omega_p \in \Omega_p A$, $\omega_q \in \Omega_q A$, $p,q \geq 0$.

Hence we get an exact sequence of cochain complexes

$$0 \longrightarrow ([\Omega_* A,\Omega_* A],d) \longrightarrow \Omega(A) \longrightarrow (\Lambda\Omega(A),d) \longrightarrow 0$$

$(\Lambda\Omega(A),d)$ is called the <u>de Rham complex</u> of (noncommutative exterior) differential forms on the unital associative k-algebra A.

Note that in general $[\Omega_* A,\Omega_* A]$ is not a graded <u>ideal</u> of $\Omega_* A$, hence the projection $\Omega(A) \to \Lambda\Omega(A)$ is not a homomorphism of graded k-algebras. For example, take any unital associative k-algebra A such that $[A,A]$ is not a two-sided ideal of A (matrix algebras will already do). Since

$[A,A] = [\Omega_* A,\Omega_* A] \cap \Omega_o A$, $[\Omega_* A,\Omega_* A]$ cannot be an ideal of $\Omega_* A$.

<u>Remark 1.1.8</u> Let now A be a unital commutative k-algebra. Then

$\Lambda^o\Omega(A) = A/[A,A] = A$, and $\Lambda^1\Omega(A) = \Omega_1 A/[A,\Omega_1 A]$

identifies with $\Omega^1_{A/k}$, the A-module of (Kähler) k-differentials for A

(cf. [Ma, pp.180-189]):
First, $[A,\Omega_1 A]$ is an A-A submodule of $\Omega_1 A$, and thus $\Lambda^1\Omega(A) = \Omega_1 A/[A,\Omega_1 A]$ becomes a symmetric A-A bimodule (i.e. can be treated as a mere left A-module). $d^o: A \to \Lambda^1\Omega(A)$ is now a universal k-derivation with the following property: For every left A-module (symmetric A-A bimodule) M, and for every k-derivation $d: A \to M$ there is a unique factorization

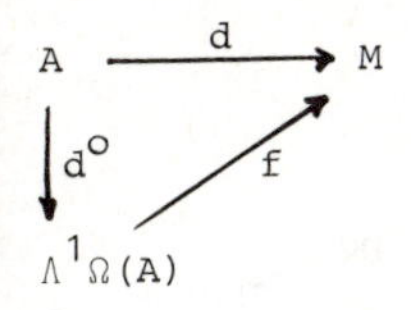

with $f \in \mathrm{Hom}_A(\Lambda^1\Omega(A),M)$.
This follows immediately from the universal factorization property of $d^o: A \to \Omega_1 A$, together with the fact that an A-A homomorphism $f: \Omega_1 A \to M$ with values in a symmetric A-A bimodule M must vanish on $[A,\Omega_1 A]$, hence factor through $\Omega_1 A \to \Lambda^1\Omega(A)$. $\Lambda^1\Omega(A)$ has thus precisely the universal property of $\Omega^1_{A/k}$; consequently, we may identify both A-modules.
Consider now the Kähler-de Rham complex $(\Omega_{A/k},d)$ of (exterior) k-differential forms on A (cf. [Bou,AX. 43]). $\Omega_{A/k} = \bigoplus_{n\geq 0} \Omega^n_{A/k} = \bigoplus_{n\geq 0} \Lambda^n\Omega^1_{A/k}$ is the exterior algebra of the A-module $\Omega^1_{A/k}$, together with the "outer derivative on forms" $d: \Omega_{A/k} \to \Omega_{A/k}$, which gives $\Omega_{A/k}$ the structure of a skew-commutative d.g. (cochain) algebra. By the universal property of $\Omega(A) = (\Omega_* A,d)$ we get a surjection of d.g. (cochain) algebras $\pi: \Omega(A) \to \Omega_{A/k}$, which vanishes on $[\Omega_* A,\Omega_* A]$, hence induces an epimorphism of cochain complexes $\pi: \Lambda\Omega(A) \to \Omega_{A/k}$; in degree 0 and 1 this is obviously an isomorphism. As to higher degrees, it is easily seen that π is an isomorphism of cochain complexes if and only if $[\Omega_* A,\Omega_* A]$ is a two-sided ideal of $\Omega_* A$.
For $A = k[T]$, a polynomial ring in one variable over k, this simply means that $[\Omega_* A,\Omega_* A]$ contains $\bigoplus_{n\geq 2} \Omega_n A$ (since $\Omega_{A/k} = A \oplus A\varepsilon$, the ring of dual numbers over A). In a moment (1.1.13) we shall see that $[T,dT]dT \in \Omega_2 A$ is not a k-linear combination of graded commutators.

<u>Definition 1.1.9</u> of (noncommutative) <u>de Rham cohomology</u> $H^*(\Lambda\Omega(A),d)$:

$$H^n(\Lambda\Omega(A),d) = \mathrm{Ker}(\Lambda^n\Omega(A) \xrightarrow{d^n} \Lambda^{n+1}\Omega(A))/\mathrm{Im}(\Lambda^{n-1}\Omega(A) \xrightarrow{d^{n-1}} \Lambda^n\Omega(A))$$

<u>Complement 1.1.10</u> on graded traces.

Let us consider the dual complex $(\Lambda\Omega(A)^t, d^t)$ of the de Rham complex:

$$0 \longleftarrow (\Lambda^0\Omega(A))^* \xleftarrow{d^t} (\Lambda^1\Omega(A))^* \xleftarrow{d^t} (\Lambda^2\Omega(A))^* \longleftarrow \cdots$$

(where $(\ )^*$ stands for k-dual $\mathrm{Hom}_k(-,k)$), and where $d^t = \mathrm{Hom}(d,1)$).

We have

$$(\Lambda^n\Omega(A))^* = \mathrm{Hom}_k(\Omega_n A / \sum_{p+q=n} [\Omega_p A, \Omega_q A], k)$$

$$= \{\tau: \Omega_n A \to k: \ \tau \ \text{k-linear and} \ \tau(\omega_p\omega_q) = (-1)^{pq}\tau(\omega_q\omega_p)$$

$$\text{for all} \ \omega_p \in \Omega_p A, \ \omega_q \in \Omega_q A, \ p+q = n\}$$

Such k-linear $\tau: \Omega_n A \to k$ (which vanish on graded commutators) will be called graded n-traces on A (or $\Omega(A)$).

$(\Lambda\Omega(A)^t, d^t)$ is thus the chain complex of graded traces on A (on $\Omega(A)$).
Note that we have for the k-module of n-cycles:

$$Z_n(\Lambda\Omega(A)^t, d^t) = \{\tau: \Omega_n A \to k, \ \ \tau \ \text{is a closed graded n-trace on} \ A\}$$

(where closed means classically: $\tau(d\omega_{n-1}) = 0$ for all $\omega_{n-1} \in \Omega_{n-1}A$)

We can summarize:
The de Rham cohomology $H^*(\Lambda\Omega(A), d)$ measures the existence (and the amount) of nontrivial closed (noncommutative) differential forms on A.
The de Rham homology $H_*(\Lambda\Omega(A)^t, d^t)$ describes the existence (and the amount) of nontrivial closed graded traces (integrals) on A.

<u>Remark 1.1.11</u> The operator $\beta: \Omega_* A \to \Omega_* A$. We want to imitate the Hochschild boundary operator b as an operator on $\Omega_* A$.
Define $\beta(A) = 0$, and $\beta: \Omega_n A \to \Omega_{n-1}A$ $(n \geq 1)$ by the formula:

$$\beta(\omega da) = (-1)^{|\omega|}[\omega, a] = (-1)^{n-1}(\omega a - a\omega) \quad \text{for} \quad \omega \in \Omega_{n-1}A, \ a \in A.$$

We have first to make sure that β is well-defined. Note that $\Omega_n A$ is a right A^e-module (cf. the beginning of I.2.1): $\omega_n(x \otimes y^o) = y\omega_n x$
Consider now $\Omega_1 A$ as a subset of $A^e = A \otimes A^{op}$, and define a composition

$$\Omega_{n-1}A \times \Omega_1 A \rightarrow \Omega_{n-1}A$$

$$(\omega_{n-1},\omega_1) \rightarrow (-1)^n \omega_{n-1}\cdot\omega_1$$

(where ω_1 operates as an elements of A^e).

It is immediate that $(\omega_{n-1}a)\cdot\omega_1 = \omega_{n-1}\cdot(a\omega_1)$, hence we can factor through $\Omega_n A = \Omega_{n-1}A \underset{A}{\otimes} \Omega_1 A$; we get $\beta\colon \Omega_n A \rightarrow \Omega_{n-1}A$, and we have actually

$$\beta(\omega_{n-1}da) = (-1)^n \omega_{n-1}\cdot(1 \otimes a - a \otimes 1) = (-1)^n (a\omega_{n-1} - \omega_{n-1}a)$$

$$= (-1)^{n-1}[\omega_{n-1},a]$$

as desired.

By induction on n one sees easily:

$$\beta(a_o da_1 \ldots da_n) = \sum_{i=0}^{n-1} (-1)^i a_o da_1 \ldots d(a_i a_{i+1}) \ldots da_n + (-1)^n a_n a_o da_1 \ldots da_{n-1}$$

Thus we have got a representation of the Hochschild boundary operator b as an operator on the differential envelope.

Note the trivial fact: $\beta\Omega_* A \subset [\Omega_* A, \Omega_* A]$,

more precisely: $\beta\Omega_n A = [A, \Omega_{n-1}A]$, $n \geq 1$

Thus we have a surjection of graded k-modules

$$\Omega_* A/\beta\Omega_{*+1}A \rightarrow \Lambda\Omega(A)$$

which is simply given by

$$\Omega_n A/[A,\Omega_n A] \rightarrow \Omega_n A/\sum_{p+q=n} [\Omega_p A, \Omega_q A], \quad n \geq 0$$

<u>Lemma 1.1.12</u> (Lifting of $\Lambda^n\Omega(A)$ inside $\Omega_n A/\beta\Omega_{n+1}A$).

(1) The cyclic group $G_n = \langle t \rangle$ of order n operates on $\Omega_n A/\beta\Omega_{n+1}A$ via

$$t.\omega_1 \otimes \ldots \otimes \omega_n = (-1)^{n-1}\omega_n \otimes \omega_1 \otimes \ldots \otimes \omega_{n-1}, \quad \omega_i \in \Omega_1 A$$

(2) Assume $\mathbb{Q} \subset k$. Then $\mathrm{Ker}(1-t)$ maps isomorphically onto

$$\Lambda^n\Omega(A) = \Omega_n A/\sum_{p+q=n} [\Omega_p A, \Omega_q A].$$

Proof.

(1) Note first (since we are working with tensor products over A) that the action of G_n on $\Omega_n A$ is not well-defined. We have to divide out precisely $\beta\Omega_{n+1}A = [A,\Omega_n A]$ in order to get a well-defined action. We shall altogether suppress equivalence class notation.

(2) With $K_n = \sum_{p+q=n} [\Omega_p A,\Omega_q A] \bmod \beta\Omega_{n+1}A$ and $D = (1-t)$ (as usual), it is immediate that $D(\Omega_n A/\beta\Omega_{n+1}A) \subset K_n$.

We want to show that $K_n = \mathrm{Im}\, D$.

Consider $\omega_1\ldots\omega_p\bar{\omega}_1\ldots\bar{\omega}_q - (-1)^{pq}\bar{\omega}_1\ldots\bar{\omega}_q\omega_1\ldots\omega_p$ with $\omega_i,\bar{\omega}_j \in \Omega_1 A$, $1 \le i \le p$, $1 \le j \le q$.

We have (everything taken mod $\beta\Omega_{n+1}A$, of course):

$$t^q(\omega_1\ldots\omega_p\bar{\omega}_1\ldots\bar{\omega}_q) = (-1)^{(p+q-1)q}\bar{\omega}_1\ldots\bar{\omega}_q\omega_1\ldots\omega_p$$

$$= (-1)^{pq}\bar{\omega}_1\ldots\bar{\omega}_q\omega_1\ldots\omega_p$$

and consequently

$$\omega_1\ldots\omega_p\bar{\omega}_1\ldots\bar{\omega}_q - (-1)^{pq}\bar{\omega}_1\ldots\bar{\omega}_q\omega_1\ldots\omega_p$$

$$= (1-t^q)(\omega_1\ldots\omega_p\bar{\omega}_1\ldots\bar{\omega}_q)$$

$$= D(1+t+\ldots+t^{q-1})(\omega_1\ldots\omega_p\bar{\omega}_1\ldots\bar{\omega}_q) \in \mathrm{Im}\, D.$$

Since we have assumed that $\mathbb{Q} \subset k$, we get

$\Omega_n A/\beta\Omega_{n+1}A = \mathrm{Ker}(1-t) \oplus \mathrm{Im}(1-t)$, i.e.

$\mathrm{Ker}(1-t)$ maps k-isomorphically onto $\Lambda^n\Omega(A)$.

Consequence 1.1.13 Assume $\mathbb{Q} \subset k$, and let $A = k[T]$ be the polynomial ring in one variable over k. We want to show that the noncommutative de Rham complex $\Lambda\Omega(A)$ does not coincide with the usual de Rham complex $\Omega_{A/k}$.

It suffices to show that $\Lambda^2\Omega(A) \neq 0$.

Taking in account 1.1.12, we shall consider $\omega = [T,dT]dT \in \Omega_2 A$, and show that

(i) $\omega \notin \beta\Omega_3 A$

(ii) $\omega \in \mathrm{Ker}(1-t)$, where $t: \Omega_2 A/\beta\Omega_3 A \to \Omega_2 A/\beta\Omega_3 A$ is given by $t(\omega_1\omega_2) = -\omega_2\omega_1$, $\omega_1,\omega_2 \in \Omega_1 A$.

(i): We show that $\beta\omega \neq 0$.

But $\beta\omega = T[T,dT] - [T,dT]T$

$$= T^2 dT - 2TdT.T + dT.T^2 \in \Omega_1 A.$$

Indentify $A \otimes A$ with $k[X,Y]$, where $X = T \otimes 1$, $Y = 1 \otimes T$.
Then $dT = Y - X$, and left T-action is multiplication with X, whereas right T-action is multiplication with Y.

Thus: $\beta\omega = (dT)^3 = (Y-X)^3 \neq 0$

(the multiplication is now in $A \otimes A$, not in $\Omega(A)$!)

(ii): We have: $\omega = \omega_1\omega_2 - \omega_2\omega_1$

where $\omega_1 = TdT$, $\omega_2 = dT$.

Hence: $\omega = (1+t)\omega_1\omega_2$, and consequently $(1-t)\omega=0$; ω represents a non-zero element of $\Lambda^2\Omega(A)$, which shows our claim.

Remark 1.1.14 We want to consider more closely the case of an augmented k-algebra $A = k \oplus \bar{A}$, where $\bar{A}$ is the augmentation ideal of A. Then $\Omega(A)$ has an alternative description, as follows (cf. [Co, p.99]): Set $\tilde{\Omega}_n A = A \otimes \bar{A}^n$ (the tensor products are over k). We have a right $\bar{A}$-action defined by the formula

$$(a_o;a_1,\ldots,a_n).a = \sum_{i=0}^{n-1} (-1)^{n-i}(a_o;a_1,\ldots,a_i a_{i+1},\ldots,a_n,a) + (a_o;a_1,\ldots,a_{n-1},a_n a)$$

This $\bar{A}$-action is associative, and extends to a unitary right action of A on $\tilde{\Omega}_n A$, which becomes thus a right A-module.

Define a composition

$\tilde{\Omega}_m A \times \tilde{\Omega}_n A \to \tilde{\Omega}_{m+n} A$

by the formula: $\omega_m.(a_o;a_1,\ldots,a_n) = (\omega_m.a) \otimes a_1 \otimes\ldots\otimes a_n$

(with the obvious identifications).

$\tilde{\Omega}_*A = \bigoplus_{n\geq 0} \tilde{\Omega}_n A$ becomes a unital associative graded k-algebra.

$d: \tilde{\Omega}_*A \to \tilde{\Omega}_*A$, defined simply by $d(a_o;a_1,\dots,a_n) = (1;a_o,\dots,a_n)$ satisfies trivially $d^2 = 0$, and one checks that

$$d(\omega_p\omega_q) = d\omega_p\cdot\omega_q + (-1)^p\omega_p d\omega_q \qquad \text{for } \omega_p \in \tilde{\Omega}_pA,\ \omega_q \in \tilde{\Omega}_qA$$

Consequently, $\tilde{\Omega}(A) = (\tilde{\Omega}_*A,d)$ is a d.g. (cochain) algebra.

The important fact is that $\tilde{\Omega}(A)$ has the same universal property as $\Omega(A)$ (cf. 1.1.5). Hence there is an isomorphism of d.g. algebras

$$\Omega(A) \xrightarrow{\theta} \tilde{\Omega}(A)$$

which identifies $a_o da_1\dots da_n$ with $(a_o;a_1,\dots,a_n)$ (you may think of $a_1,\dots,a_n$ as elements of A or of $\bar{A}$; the notation will be coherent in either case).

Note that this isomorphism shows in particular (together with the definition of d on the $\tilde{\Omega}(A)$-side) that $\Omega(A) = (\Omega_*A,d)$ is acyclic for an augmented k-algebra $A = k \oplus \bar{A}$.

Lemma 1.1.15 Let $A = k \oplus \bar{A}$ be an augmented k-algebra. For $n \geq 0$ consider the isomorphism

$$\theta:\ \Omega_nA \ni a_o da_1\dots da_n \to (a_o;a_1,\dots,a_n) \in A \otimes \bar{A}^n$$

Then the image of a graded commutator is given by the formula:

$$\theta([a_o da_1\dots da_K,\ a_{K+1}da_{K+2}\dots da_{n+1}])$$
$$= \sum_{i=0}^{K}(-1)^{K-i}(a_o;\dots a_i a_{i+1}\dots a_{n+1})$$
$$- \sum_{i=0}^{n-K}(-1)^{K(n-K)+n-K-i}(a_{K+1};\dots a_{K+1+i}a_{K+1+i+1}\dots a_K)$$

(Convention: $a_{n+2} = a_o$)

Proof. It is immediate that

$$\theta((a_o da_1 \ldots da_K)(a_{K+1} da_{K+2} \ldots da_{n+1})) = \sum_{i=0}^{K} (-1)^{K-i} (a_o; \ldots a_i a_{i+1} \ldots a_{n+1})$$

and that

$$\theta((a_{K+1} da_{K+2} \ldots da_{n+1})(a_o da_1 \ldots da_K))$$

$$= \sum_{i=0}^{n-K} (-1)^{n-K-i} (a_{K+1}; \ldots a_{K+1+i} a_{K+1+i+1}, \ldots a_K)$$

which gives our assertion.

Consequence 1.1.16 In the situation 1.1.15, assume furthermore that $a_o, a_1, \ldots, a_{n+1} \in \bar{A}$. Consider our standard operation of $G_{n+1} = \langle t \rangle$ on $\bar{A}^{n+1} = \bar{A} \otimes \bar{A}^n \subset A \otimes \bar{A}^n$. Then

$$\theta([a_o da_1 \ldots da_K, a_{K+1} da_{K+2} \ldots da_{n+1}]) = (-1)^K b(a_o; \ldots a_{n+1}) \bmod \mathrm{Im}(1-t).$$

Proof. We have

$$t^{K+1}(a_{K+1}; \ldots a_{K+1+i} a_{K+1+i+1}, \ldots a_K)$$

$$= \begin{cases} (-1)^{(K+1)n} (a_o; \ldots a_{K+1+i} a_{K+1+i+1}, \ldots a_{n+1}) & \text{for } i < n-K \\ (-1)^{(K+1)n} (a_{n+1} a_o; a_1, \ldots, a_n) & \text{for } i = n-K \end{cases}$$

Thus we obtain

$$\sum_{i=0}^{n-K} (-1)^{K(n-K)+n-K-i} (a_{K+1}; \ldots a_{K+1+i} a_{K+1+i+1}, \ldots a_K)$$

$$= \sum_{i=0}^{n-K-1} (-1)^i t^{n-K} (a_o; \ldots a_{K+1+i} a_{k+1+i+1}, \ldots a_{n+1}) + (-1)^{n-K} t^{n-K} (a_{n+1} a_o; a_1, \ldots a_n)$$

$$= - \sum_{i=K+1}^{n} (-1)^{K-i} t^{n-K} (a_o; \ldots a_i a_{i+1}, \ldots a_{n+1}) + (-1)^{n-K} t^{n-K} (a_{n+1} a_o; a_1, \ldots a_n)$$

From 1.1.15 we obtain finally

$$\theta([a_o da_1 \ldots da_K, a_{K+1} da_{K+2} \ldots da_{n+1}])$$

$$= (-1)^K b(a_o; a_1, \ldots, a_{n+1}) - (1-t^{n-K}) \Big(\sum_{i=K+1}^{n} (-1)^{K-i} (a_o; \ldots a_i a_{i+1}, \ldots a_{n+1})$$

$$+ (-1)^{n+1} (a_{n+1} a_o; a_1, \ldots a_n) \Big)$$

which proves our claim.

Lemma 1.1.17 Let $A = k \oplus \bar{A}$ be an augmented k-algebra. Then, for every $n \geq 1$, the isomorphism

$$\theta: \Omega_n A \ni a_o da_1 \ldots da_n \rightarrow (a_o; a_1, \ldots, a_n) \in A \otimes \bar{A}^n$$

induces an isomorphism

$$\theta: \Lambda^n \Omega(A)/d\Lambda^{n-1}\Omega(A) \rightarrow [\bar{A}^{n+1}/(1-t)] \text{mod Im } b.$$

Proof. In complement to out result 1.1.16 we have the formula

$$(*) \quad a_o da_1 \ldots da_n - (-1)^n a_n da_o \ldots da_{n-1}$$
$$= (-1)^n [da_o \ldots da_{n-1}, a_n] - \sum_{i=0}^{n-1} (-1)^i da_o \ldots d(a_i a_{i+1}) \ldots da_n$$

Now, by 1.1.16, θ is actually well-defined as an application from

$\Lambda^n \Omega(A)/d\Lambda^{n-1}\Omega(A)$ to $[\bar{A}^{n+1}/(1-t)]$mod Im b.

But, by virtue of the formula (*), combined with 1.1.16 (read in the other direction), the inverse application θ^{-1} is well-defined, too. This shows our assertion.

Theorem 1.1.18 Assume $\mathbb{Q} \subset k$, and let $A = k \oplus \bar{A}$ be an augmented k-algebra.
Then, for every $n \geq 1$, we have an exact sequence

$$0 \longrightarrow H^n(\Lambda\Omega(A), d) \xrightarrow{\theta} \overline{HC}_n(A) \xrightarrow{B} \bar{H}_{n+1}(A)$$

(i.e. non-commutative de Rham cohomology lies inside reduced cyclic homology).

Proof. In order to show the injectivity (and well-definedness!) of $H^n(\Lambda\Omega(A), d) \xrightarrow{\theta} \overline{HC}_n(A)$, we have only to exhibit a commutative diagram

$$\begin{array}{ccc}
\Lambda^n \Omega(A)/d\Lambda^{n-1}\Omega(A) & \xrightarrow{\theta} & [\bar{A}^{n+1}/(1-t)]\text{mod Im } b \\
\downarrow d & & \downarrow b \\
\Lambda^{n+1}\Omega(A) & \xrightarrow{\varphi} & \bar{A}^n/(1-t)
\end{array}$$

which, by I.2.5.18 and 1.1.17, will show our first claim.
Now, for every upper triangular matrix $\varepsilon = (\varepsilon_{ij})_{0\le i\le j\le n+1}$ with entries $\varepsilon_{ij} \in \{0,+1,-1\}$ (and zeroes on the main diagonal) you can define

$\varphi_\varepsilon : \Omega_{n+1}A \to A \otimes \bar{A}^{n-1}$ by the formula

$$\varphi_\varepsilon(a_0 da_1 \dots da_{n+1}) = \sum_{i<j} \varepsilon_{ij}(a_i a_{i+1}; \dots a_j a_{j+1}, \dots a_{i-1})$$
$$+ (-1)^n (a_{n+1}a_0 a_1; a_2, \dots a_n)$$

In order to get a factorization $\varphi: \Lambda^{n+1}\Omega(A) \to \bar{A}^n/(1-t)$ which makes our diagram commutative, there must be further conditions on the ε_{ij}. One checks that the nonzero ε_{ij} have to be defined for $(i,j) \neq (0,n+1)$, $0 \le i \le n-1$, $2 \le j \le n+1$; $i < j-1$ such that

$$\varepsilon_{ij} = (-1)^{n+1}\varepsilon_{i+1,j+1} \quad (j < n+1).$$

This establishes finally the inclusion of $H^n(\Lambda\Omega(A),d)$ into $\overline{HC}_n(A)$, for $n \ge 1$.
We are let to prove the equality $\theta(\mathrm{Ker}\ d) = \mathrm{Ker}\ B$. Denote by $\mathbb{B}$ the operator on $\Omega_* A$ which corresponds to B on $\bar{C}(A)$: $\mathbb{B} = \theta^{-1}B\theta : \Omega_n A \to \Omega_{n+1}A$ is clearly given by the formula

$$\mathbb{B}(a_0 da_1 \dots da_n) = \sum_{i=0}^{n} (-1)^{in} da_i \dots da_n da_0 \dots da_{i-1}.$$

The equality $Bb + bB = 0$ on the $A\otimes\bar{A}^*$-side correspond to $\mathbb{B}\beta + \beta\mathbb{B} = 0$ on the $\Omega_* A$-side, hence we get an induced operator

$$\mathbb{B}: \Omega_n A/\beta\Omega_{n+1}A \to \Omega_{n+1}A/\beta\Omega_{n+2}A.$$

Now, according to 1.1.12, $\Lambda^n\Omega(A)$ can be identified with $\mathrm{Ker}(1-t)$ (for the signed cyclic permutation operator t on $\Omega_n A/[A,\Omega_n A]$).
We obtain immediately

$$\mathbb{B}(a_0 da_1 \dots da_n) = da_0 da_1 \dots da_n + t(da_0 da_1 \dots da_n) + \dots + t^n(da_0 da_1 \dots da_n)$$
$$= Nd(a_0 da_1 \dots da_n)$$

(where $N = 1+t+\dots+t^n$ is the norm operator on $\Omega_{n+1}A/\beta\Omega_{n+2}A$).
Note that N is, up to a scalar factor, projection on $\mathrm{Ker}(1-t) = \Lambda^{n+1}\Omega(A)$.

Since $d[\Omega_*A,\Omega_*A] \subset [\Omega_*A,\Omega_*A]$ (i.e. d maps the complementary k-submodules of the various $\mathrm{Ker}(1-t)$ into each other), we get finally what we want:
First, for $\omega \in \mathrm{Ker}(1-t) = \Lambda^n\Omega(A)$ we have $\mathbb{B}(\omega) = Nd(\omega) \in \mathrm{Ker}(1-t)$ $= \Lambda^{n+1}\Omega(A)$, hence $\mathbb{B}(\omega) = 0$ if and only if $d(\omega) \in [\Omega_*A,\Omega_*A]_{n+1}/\beta\Omega_{n+2}A$, i.e. if and only if $\omega \in \mathrm{Ker}(\Lambda^n\Omega(A) \xrightarrow{d} \Lambda^{n+1}\Omega(A))$. This finishes the proof of our theorem.

Corollary 1.1.19 (Noncommutative Poincaré Lemma)
Assume $\mathbb{Q} \subset k$, and let V be a flat k-module. Consider $A = T(V)$, the tensor algebra of V over k. Then

$$H^n(\Lambda\Omega(A),d) = 0 \quad \text{for} \quad n \geq 1.$$

Proof. 1.1.18 together with I.2.5.13: $\overline{HC}_n(A) = 0$ for $n \geq 1$.

II.1.2 Cyclic homology and de Rham cohomology of commutative algebras.

Let A be a unital commutative k-algebra. The Hochschild homology $H_*(A)$ of A has a particular nice structure of a unital skew commutative graded (associative) A-algebra. We shall first discuss this structure, then its relation to the Kähler-de Rham algebra of k-differential forms on A, and finally look at a refinement of 1.1.18: for a sufficiently large class of commutative algebras we obtain, in characteristic zero, a decomposition of cyclic homology in terms of de Rham cohomology.

Remark 1.2.1 Consider the symmetric group γ_n consisting of all permutations of the set $\{1,2,\ldots,n\}$.
Fix (p,q) such that $p+q = n$, $1 \leq p \leq n-1$.
A (p,q)-shuffle σ is an element of γ_n such that

$$\sigma(1) <\ldots< \sigma(p) \quad \text{and} \quad \sigma(p+1) <\ldots< \sigma(p+q).$$

Visualization:

Take a deck of p ordered red cards and a deck of q ordered blue cards. A (p,q)-shuffle describes one possible way of shuffling the deck of p red cards through the deck of q blue cards such that the internal order of either deck is preserved. $\sigma(i)$, $1 \leq i \leq p$, will indicate the position of the i-th red card, $\sigma(p+j)$, $1 \leq j \leq q$, the position of the j-th blue card.
$\gamma_{p,q}$ shall denote the set of (p,q)-shuffles (located in γ_{p+q}). Note

that card $\gamma_{p,q} = \frac{(p+q)!}{p!q!}$.

In order to get accustomed to shuffle-arguments, consider now $\gamma_{q,p} \subset \gamma_{p+q}$, the set of (q,p)-shuffles.
It has the same cardinality as $\gamma_{p,q}$.
Define the permutation $\alpha \in \gamma_{p+q}$ by

$$\alpha(1) = q+1, \ldots \alpha(p) = q + p, \quad \alpha(p+1) = 1, \ldots \alpha(p+q) = q$$

(interchange the [1,...p]-segment and the [p+1,...p+q]-segment).

$\alpha \in \gamma_{p,q}$, $\alpha^{-1} \in \gamma_{q,p}$.

The sign $\varepsilon(\alpha)$ of α is given by $\varepsilon(\alpha) = (-1)^{pq}$ (you have to do p transpositions q times).
The application

$$\alpha^*: \gamma_{q,p} \ni \tau \longrightarrow \tau \circ \alpha \in \gamma_{p,q}$$

is a bijection between $\gamma_{q,p}$ and $\gamma_{p,q}$
(its inverse is induced by α^{-1} via the same formula).

Remark 1.2.2 Let k be a commutative ring, V a k-module. Denote by V^m the m-fold tensor product over k.
Define a composition

$$V^p \times V^q \longrightarrow V^{p+q}$$

$$((a_1,\ldots,a_p),(a_{p+1},\ldots,a_{p+q})) \longrightarrow [a_1,\ldots,a_p;a_{p+1},\ldots,a_{p+q}]$$

$$:= \sum_{\sigma\in\gamma_{p,q}} \varepsilon(\sigma)(a_{\sigma^{-1}(1)},\ldots,a_{\sigma^{-1}(p+q)})$$

(extended by bilinearily, of course).

This "shuffle-product" is

(1) associative, i.e. we have (in V^{p+q+r}):

$$[[a_1,\ldots,a_p;a_{p+1},\ldots,a_{p+q}];a_{p+q+1},\ldots,a_{p+q+r}]$$

$$= [a_1,\ldots,a_p;[a_{p+1},\ldots,a_{p+q};a_{p+q+1},\ldots,a_{p+q+r}]]$$

(2) Skew-commutative, i.e. the following equality holds:

$$[a_1,\dots,a_p;a_{p+1},\dots,a_{p+q}] = (-1)^{pq}[a_{p+1},\dots,a_{p+q};a_1,\dots,a_p]$$

ad(1): Associativity is clear, since both sides of our (claimed) equality are given by

$$\sum_{\sigma\in\gamma_{p,q,r}} \varepsilon(\sigma)(a_{\sigma^{-1}(1)},\dots,a_{\sigma^{-1}(p+q)})$$

where $\gamma_{p,q,r} \subset \gamma_{p+q+r}$ is the subset of those permutations σ of $\{1,\dots p,\dots p+q,\dots p+q+r\}$ such that

$$\sigma(1) <\dots< \sigma(p) \quad\text{and}\quad \sigma(p+1) <\dots< \sigma(p+q) \quad\text{and}\quad \sigma(p+q+1) <\dots< \sigma(p+q+r)$$

(think of three decks of cards of three different colours and of the various "associative procedures" to obtain all shuffles which preserve the inner order of either deck).

ad(2):

$$[a_1,\dots,a_p;a_{p+1},\dots,a_{p+q}] = \sum_{\sigma\in\gamma_{p,q}} \varepsilon(\sigma)(a_{\sigma^{-1}(1)},\dots,a_{\sigma^{-1}(p+q)})$$

$$[a_{p+1},\dots,a_{p+q};a_1,\dots,a_p] = [a_{\alpha^{-1}(1)},\dots,a_{\alpha^{-1}(q)};a_{\alpha^{-1}(q+1)},\dots,a_{\alpha^{-1}(p+q)}]$$

(where α is as in 1.2.1)

Thus:

$$[a_{p+1},\dots,a_{p+q};a_1,\dots,a_p] = \sum_{\tau\in\gamma_{q,p}} \varepsilon(\tau)(a_{\alpha^{-1}\tau^{-1}(1)},\dots,a_{\alpha^{-1}\tau^{-1}(p+q)})$$

But we have already seen in 1.2.1 that $\tau \to \tau\circ\alpha$ establishes a bijection between $\gamma_{q,p}$ and $\gamma_{p,q}$, and that $\varepsilon(\alpha) = (-1)^{pq}$.

Hence

$$[a_1,\dots,a_p;a_{p+1},\dots,a_{p+q}] = \sum_{\sigma\in\gamma_{p,q}} \varepsilon(\sigma)(a_{\sigma^{-1}(1)},\dots,a_{\sigma^{-1}(p+q)})$$

$$= \sum_{\tau\in\gamma_{q,p}} \varepsilon(\tau\circ\alpha)(a_{(\tau\circ\alpha)^{-1}(1)},\dots,a_{(\tau\circ\alpha)^{-1}(p+q)})$$

$$= (-1)^{pq} \sum_{\tau \in \gamma_{q,p}} \varepsilon(\tau)(a_{\alpha^{-1}\tau^{-1}(1)},\dots,a_{\alpha^{-1}\tau^{-1}(p+q)})$$

$$= (-1)^{pq}[a_{p+1},\dots,a_{p+q};a_1,\dots,a_p],$$

as claimed.

Remark-Definition 1.2.3 Let k be a commutative ring, and let A be a commutative unital k-algebra. Consider the acyclic Hochschild complex with its original correct grading, i.e. in the form (A^{*+2},b') - such that the Hochschild complex (A^{*+1},b) identifies with $(A \otimes_{A^e} A^{*+2}, 1 \otimes b')$: cf. I.2.1.2(2).

Notation: $S(A) = (A^{*+2},b')$

$$S_n(A) = A \otimes A^n \otimes A, \quad n \geq 0.$$

In particular: $S_o(A) = A \otimes A = A^e$.

Define the shuffle-product on $S(A)$:

$$S_p(A) \times S_q(A) \longrightarrow S_{p+q}(A)$$

$$(a_o \otimes (a_1,\dots,a_p) \otimes a_{p+1}, b_o \otimes (b_1,\dots,b_q) \otimes b_{q+1}) \longmapsto$$

$$(a_o b_o) \otimes [a_1,\dots,a_p;b_1,\dots,b_q] \otimes (a_{p+1}b_{q+1})$$

This product is $A\otimes A$-bilinear, hence converts $S(A)$ into a skew-commtative graded (associative) $A\otimes A$-algebra, by the properties (1) and (2) of 1.2.2.

Proposition 1.2.4 $(S(A),b')$ is a d.g. (chain) algebra, i.e. b' is a graded derivation:
Let x be homogeneous of degree p, y arbitrary, then

$$b'(x.y) = b'(x).y + (-1)^p x.b'(y).$$

Proof. Clearly we may assume y to be homogeneous (of degree q, say), too.
We shall prove the graded derivation property by induction on $n = p+q$. Let us begin with $n = 1$. By virtue of skew-commutativity, we may assume $x \in S_o(A)$, $y \in S_1(A)$.

$$X = a_o \otimes a_1, \quad y = b_o \otimes b_1 \otimes b_2$$

$$x.y = a_o b_o \otimes b_1 \otimes a_1 b_2$$

$$b'(x.y) = a_o b_o b_1 \otimes a_1 b_2 - a_o b_o \otimes b_1 a_1 b_2$$

$$b'(x).y = 0$$

$$x.b'(y) = (a_o \otimes a_1)(b_o b_1 \otimes b_2 - b_o \otimes b_1 b_2)$$

$$= a_o b_o b_1 \otimes a_1 b_2 - a_o b_o \otimes a_1 b_1 b_2$$

and we have shown our equality.
As to the inductive step, we shall need a technical lemma.

Consider $x = 1 \otimes (a_1,\ldots,a_p) \otimes 1, \quad y = 1 \otimes (b_1,\ldots,b_q) \otimes 1$

and $\bar{x} = 1 \otimes (a_2,\ldots,a_p) \otimes 1, \quad \bar{y} = 1 \otimes (b_2,\ldots,b_q) \otimes 1.$

Recall the contracting homotopy $s\colon S_*(A) \to S_{*+1}(A)$. We can write:

$$x = s(a_1\bar{x}) = s((a_1 \otimes 1).\bar{x})$$

$$y = s(b_1\bar{y}) = s((b_1 \otimes 1).\bar{y})$$

<u>Lemma</u>. x,y as above; then the following identities hold:

(1) $x.y = s(a_1\bar{x}.y + (-1)^p b_1 x.\bar{y})$

(2) $b'x.y = a_1\bar{x}.y - s(a_1(b'\bar{x}).y + (-1)^p b_1(b'x).\bar{y})$

(3) $x.b'y = b_1 x.\bar{y} + s(a_1\bar{x}.b'y - (-1)^p b_1 x.b'\bar{y})$

<u>Proof</u>. Note first that (2) and (3) are equivalent by skew-commutativity. We shall only prove (1), since (2) is shown by the same device. In order to be conform with our former notation, let us write

$$x = 1 \otimes (a_1,\ldots,a_p) \otimes 1, \quad y = 1 \otimes (a_{p+1},\ldots,a_{p+q}) \otimes 1$$

$$\bar{x} = 1 \otimes (a_2,\ldots,a_p) \otimes 1, \quad \bar{y} = 1 \otimes (a_{p+2},\ldots,a_{p+q}) \otimes 1$$

Thus

$$x.y = 1 \otimes [a_1,\dots,a_p;a_{p+1},\dots,a_{p+q}] \otimes 1$$

$$= \sum_{\sigma\in\gamma_{p,q}} \varepsilon(\sigma)1 \otimes (a_{\sigma^{-1}(1)},\dots,a_{\sigma^{-1}(p+q)}) \otimes 1.$$

In order to evaluate conveniently

$$s(a_1\bar{x}.y + (-1)^p a_{p+1}x.\bar{y})$$

let us observe the following:

Write $\gamma_{p,q} = \gamma^1_{p,q} \cup \gamma^2_{p,q}$

where $\gamma^1_{p,q} = \{\sigma \in \gamma_{p,q}\colon \sigma(1) = 1\}$

$\gamma^2_{p,q} = \{\sigma \in \gamma_{p,q}\colon \sigma(p+1) = 1\}$.

We can identify $\gamma^1_{p,q}$ with $\gamma_{p-1,q}$, considered as a subset of $\gamma_{\{2,\dots,p+q\}}$.

As to $\gamma^2_{p,q}$, consider the cyclic permutation $\lambda = (1,2,\dots,p+1)$. $\varepsilon(\lambda) = (-1)^p$. $\gamma^2_{p,q}$ identifies with $\gamma_{p,q-1}$, equally looked at as a subset of $\gamma_{\{2,\dots,p+q\}}$ via the bijection

$$\gamma_{\{2,..,p+q\}} \supset \gamma_{p,q-1} \ni \sigma \to \sigma\circ\lambda \in \gamma^2_{p,q}.$$

Now, let us write down explicitely

$$s(a_1\bar{x}.y+(-1)^p a_{p+1}x.\bar{y})$$

$$= \sum_{\sigma\in\gamma_{p-1,q}} \varepsilon(\sigma)1 \otimes (a_1,a_{\sigma^{-1}(2)},\dots,a_{\sigma^{-1}(p+q)}) \otimes 1$$

$$+ (-1)^p \sum_{\sigma\in\gamma_{p,q-1}} \varepsilon(\sigma)1 \otimes (a_{p+1},a_{\lambda^{-1}\sigma^{-1}(2)},\dots,a_{\lambda^{-1}\sigma^{-1}(p+q)}) \otimes 1$$

$$= \sum_{\sigma\in\gamma^1_{p,q}} \varepsilon(\sigma)1 \otimes (a_{\sigma^{-1}(1)},a_{\sigma^{-1}(2)},\dots,a_{\sigma^{-1}(p+q)}) \otimes 1$$

$$+ \sum_{\tau\in\gamma^2_{p,q}} \varepsilon(\tau)1 \otimes (a_{\tau^{-1}(1)},a_{\tau^{-1}(2)},\dots,a_{\tau^{-1}(p+q)}) \otimes 1$$

$$= \sum_{\sigma\in\gamma_{p,q}} \varepsilon(\sigma) 1 \otimes (a_{\sigma^{-1}(1)},\dots,a_{\sigma^{-1}(p+q)}) \otimes 1$$

$$= x.y$$

as claimed.

Let us return to the proof of our proposition. We have to attack the inductive step.

Write $\quad x.y = s(a_1\bar{x}.y + (-1)^p b_1 x.\bar{y})$

as guaranteed by our lemma (note that we may assume x and y to be of the particular form treated by the lemma, since b' is $A\otimes A$-linear). Then

$$\begin{aligned} b'(x.y) &= b's(a_1\bar{x}.y + (-1)^p b_1 x.\bar{y}) \\ &= a_1\bar{x}.y + (-1)^p b_1 x.\bar{y} - sb'(a_1\bar{x}.y + (-1)^p b_1 x.\bar{y}) \\ &= a_1\bar{x}.y + (-1)^p b_1 x.\bar{y} - s(a_1 b'(\bar{x}.y) + (-1)^p b_1 b'(x.\bar{y})) \end{aligned}$$

Investing our inductive hypothesis and rearranging everything, we obtain:

$$\begin{aligned} b'(x.y) = {} & a_1\bar{x}.y - s(a_1(b'\bar{x}).y) - (-1)^p s(b_1(b'x).\bar{y}) \\ & + (-1)^p b_1 x.\bar{y} + (-1)^p s(a_1\bar{x}.b'y) - s(b_1 x.b'\bar{y}) \end{aligned}$$

By (2) and (3) of our technical lemma, the first three terms on the right side are equal to $b'x.y$, and the last three terms on the right side are equal to $(-1)^p x.b'y$, which proves the asserted identity. This finishes the proof of our proposition.

<u>Corollary 1.2.5</u> Let k be a commutative ring, and let A be a unital commutative k-algebra.

Define the <u>shuffle product</u> on $C(A) = (A^{*+1}, b)$, the Hochschild complex of A, by the formula

$$(a,a_1,\dots,a_p).(a',a_{p+1},\dots,a_{p+q}) = \sum_{\sigma\in\gamma_{p,q}} \varepsilon(\sigma)(aa',a_{\sigma^{-1}(1)},\dots,a_{\sigma^{-1}(p+q)})$$

Then $C(A)$ becomes a skew-commutative d.g. (associative) A-algebra.

<u>Proof</u>. This is a straight forward consequence of 1.2.4, since $C(A) =$

(A^{*+1},b) is merely a scalar-extension of $S(A) = (A^{*+2},b')$:

$C(A) = A \underset{A\otimes A}{\otimes} S(A)$ as a d.g. algebra (with $b = 1 \otimes b'$)

Recall I.2.1.2 (2):

$$C_n(A) = A \otimes A^n = A \underset{A\otimes A}{\otimes} (A \otimes A^n \otimes A) = A \underset{A\otimes A}{\otimes} S_n(A)$$

with the identification:

$$(a_{n+1}aa_o, a_1, \ldots, a_n) \leftrightarrow a \otimes_{A^e} (a_o \otimes (a_1, \ldots, a_n) \otimes a_{n+1})$$

Complement 1.2.6 In the situation 1.2.5 consider $\overline{C}(A) = (A \otimes \overline{A}^*, b)$, the normalized Hochschild complex of our commutative algebra A, and recall the exact sequence of chain complexes

$$0 \longrightarrow (D_*, b) \longrightarrow (A^{*+1}, b) \xrightarrow{\pi} (A \otimes \overline{A}^*, b) \longrightarrow 0$$

Since D_* is clearly a graded ideal of $C(A)$ for the shuffle-product structure, $\overline{C}(A)$ inherits a natural quotient structure of a skew-commutative d.g. A-algebra, with shuffle-product given by the same formula as for $C(A)$ (up to a semi-colon, at least in our notation..).

Consequence 1.2.7 By the graded derivation property of $b: C(A) \to C(A)$ we obtain: $Z_*(A) = \operatorname{Ker} b$ is a graded subalgebra of $C(A)$ (for the shuffle-product), and $B_*(A) = \operatorname{Im} b$ is a graded ideal of $Z_*(A)$. The same holds for $\overline{Z}_*(A)$ and $\overline{B}_*(A)$ in $\overline{C}(A)$. Hence $H_*(A) = Z_*(A)/B_*(A)$ $= \overline{Z}_*(A)/\overline{B}_*(A)$ is a skew-commutative (associative) graded A-algebra ($A = H_o(A)$!) with respect to the induced shuffle-product.

Remark 1.2.8 Let A be, as always in this paragraph, a unital commutative k-algebra. We have to look a little bit more closely at $\Omega^1_{A/k}$, the A-module of (Kähler) k-differentials on A. Recall first 1.1.1:

$$\Omega_1 A = \operatorname{Ker}(A \otimes A \xrightarrow{b'} A) \quad \text{together with} \quad d^o: A \longrightarrow \Omega_1 A$$
$$a \otimes b \longrightarrow ab \qquad\qquad a \longrightarrow 1\otimes a - a\otimes 1$$

have the following universal factorization property:
For every A-A-bimodule M and for every k-derivation $d: A \to M$ there is a unique A-A-homormorphism $f: \Omega_1 A \to M$ such that $d = f\, d^o$.
The commutativity of A implies that $b': A \otimes A \to A$ is actually a

homomorphis of unital commutative k-algebras; consequently $J = \Omega_1 A =$ Ker b' is an ideal of $A \otimes A$.
Furthermore, the A-A-submodule $[A,\Omega_1 A]$ of $\Omega_1 A$ is nothing else than J^2, the square of the ideal $J = \Omega_1 A$, by virtue of the identity

$$\begin{aligned} d^o x d^o y &= (1 \otimes x - x \otimes 1)(1 \otimes y - y \otimes 1) \\ &= 1 \otimes xy - x \otimes y - y \otimes x + xy \otimes 1 \\ &= (1 \otimes y - y \otimes 1)x - x(1 \otimes y - y \otimes 1) \\ &= [d^o y, x]. \end{aligned}$$

As a consequence, $\Omega^1_{A/k} = \Omega_1 A/[A,\Omega_1 A] = J/J^2$ together with $d^o\colon A \to \Omega^1_{A/k}$ (where $d^o a = (1\otimes a - a\otimes 1) \bmod J^2$) have the following properties:

(1) $\Omega^1_{A/k}$ is a symmetric A-A-bimodule (i.e. may be treated merely as a left A-module), and $d^o\colon A \to \Omega^1_{A/k}$ is a k-derivation from A to $\Omega^1_{A/k}$.

(2) $d^o\colon A \to \Omega^1_{A/k}$ is universal in the following sense: For every left A-module (i.e. symmetric A-A-bimodule) M and for every k-derivation $d\colon A \to M$ there is a unique A-homomorphism $f\colon \Omega^1_{A/k} \to M$ such that $d = f \circ d^o$ (cf. 1.1.8).

Lemma 1.2.9 Let A be a unital commutative k-algebra. There is a commutative diagram

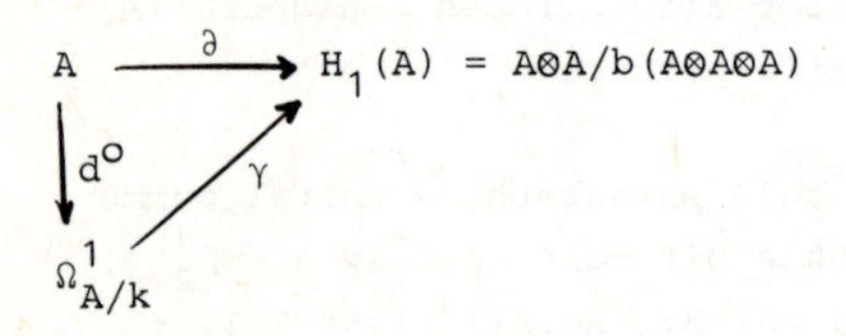

where $\partial(a) = 1\otimes a \bmod b(A\otimes A\otimes A)$ for all $a \in A$
and $\gamma(a_o d^o a_1) = a_o \otimes a_1 \bmod b(A\otimes A\otimes A)$ for all $a_o, a_1 \in A$
∂ is a k-derivation, and γ is an isomorphism of left A-modules.

Proof. First, $b(A\otimes A\otimes A)$ is the A-submodule of $A \otimes A$ spanned by all elements of the form

$$(xy,z) - (x,yz) + (xz,y), \quad x,y,z \in A.$$

This implies that $\partial: A \to H_1(A)$, defined as stated above, is a k-derivation from A to $H_1(A)$.
The universal factorization property of $d^o: A \to \Omega^1_{A/k}$ guarantees the existence and the explicit form of the A-homomorphism γ.
In order to prove that γ is an A-isomorphism, it suffices to show that $\partial: A \to H_1(A)$ has the same universal property as $d^o: A \to \Omega^1_{A/k}$.
But let $d: A \to M$ be any k-derivation of A into some left A-module M. There is an obvious factorization

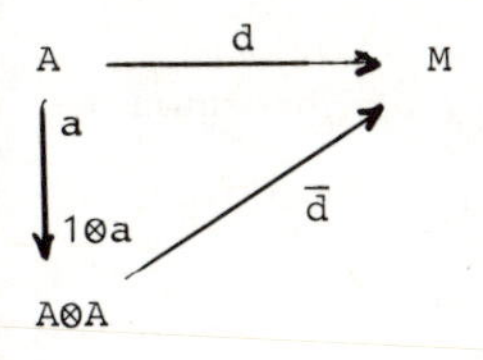

with $\bar{d}(\Sigma c_i \otimes a_i) = \Sigma c_i d(a_i)$; $\bar{d} \in \mathrm{Hom}_A(A\otimes A, M)$.
Furthermore, since d is a k-derivation, $\bar{d}$ vanishes on $b(A\otimes A\otimes A)$, i.e. we get a further factorization

$$\begin{array}{ccc} A\otimes A & \xrightarrow{\bar{d}} & M \\ \downarrow{\scriptstyle \mathrm{can}} & \nearrow{\scriptstyle f} & \\ H_1(A) & & \end{array}$$

with $f \in \mathrm{Hom}_A(H_1(A), M)$.

f is unique, since $H_1(A) = A\otimes A/b(A\otimes A\otimes A)$ is generated by $\partial(A)$ as a left A-module.

<u>Corollary 1.2.10</u> Let $(\Omega_{A/k}, d)$ be the Kähler-de Rham complex of k-differential forms on A (cf. 1.1.8).
Then there is a homomorphism of skew-commutative graded A-algebras $\gamma: \Omega_{A/k} \to H_*(A)$ which prolongs the A-isomorphism

$$\gamma = \gamma_1 : \Omega^1_{A/k} \to H_1(A)$$

<u>Proof</u>. This follows from the universal property of $\Omega_{A/k} = \bigoplus_{n\geq 0} \Lambda^n \Omega^1_{A/k}$ as the exterior algebra of the A-module $\Omega^1_{A/k}$.

<u>Proposition 1.2.11</u> For every $n \geq 0$ there is a commutative square

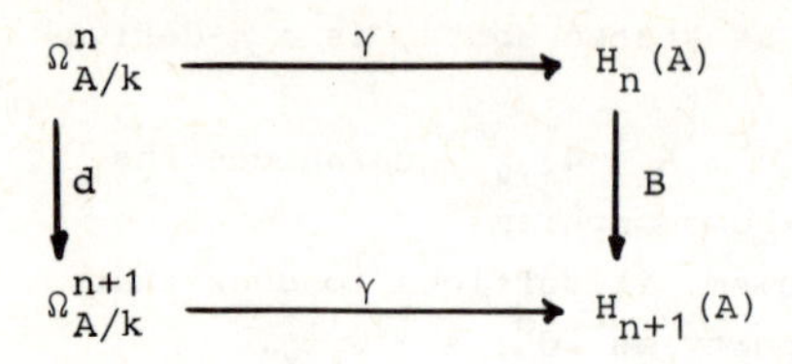

Proof. By 1.2.6 we may assume that the shuffle-product structure on $H_*(A)$ is induced by the shuffle-product structure on $\overline{C}(A)$, the normalized Hochschild complex of A.
First, recall the argument concerning the associativity of the shuffle-product (1.2.2(1)):

$$[[a_1,\ldots,a_p;a_{p+1},\ldots,a_{p+q}];\ a_{p+q+1},\ldots,a_{p+q+r}]$$

$$= \sum_{\sigma\in\gamma_{p,q,r}} \varepsilon(\sigma)(a_{\sigma^{-1}(1)},\ldots,a_{\sigma^{-1}(p+q)})$$

where $\gamma_{p,q,r}$ is the subset of γ_{p+q+r} consisting of all (p,q,r)-shuffles.
This yields immediately the following formula in $\overline{C}(A)$:

$$(a_o;a_1).(1;a_2).\ \ldots\ (1;a_n) = \sum_{\sigma\in\gamma_n} \varepsilon(\sigma)(a_o;a_{\sigma^{-1}(1)},\ldots,a_{\sigma^{-1}(n)})$$

(since γ_n is the set of all $(\underbrace{1,1,\ldots,1}_{n\text{ times}})$-shuffles in γ_n)

Now, with $\omega_n = a_o da_1 \wedge\ldots\wedge da_n \in \Omega^n_{A/k}$ we have, by the multiplicativity of γ.

$$\gamma(\omega_n) = \gamma(a_o da_1).\gamma(da_2).\ \ldots\ \gamma(da_n)$$

$$= (a_o;a_1).(1;a_n)\ldots(1;a_n)$$

$$= \sum_{\sigma\in\gamma_n} \varepsilon(\sigma)(a_o;a_{\sigma^{-1}(1)},\ldots,a_{\sigma^{-1}(n)}) \bmod \overline{B}_n(A)$$

Since $d\omega_n = da_o \wedge da_1 \wedge\ldots\wedge da_n \in \Omega^{n+1}_{A/k}$, we get analogously

$$\gamma(d\omega_n) = \sum_{\tau\in\gamma_{n+1}} \varepsilon(\tau)(1;a_{\tau^{-1}(0)},\ldots,a_{\tau^{-1}(n)}) \bmod \overline{B}_{n+1}(A).$$

Recall that $B: A\otimes\overline{A}^n \to A\otimes\overline{A}^{n+1}$ is given by the formula

$$B(a_o;a_1,\ldots,a_n) = \sum_{i=0}^{n} (-1)^{in}(1;a_i,\ldots,a_n,a_o,\ldots,a_{i-1})$$

(cf. I.2.5.2)

Let us write this formula in another way.
Consider the cyclic permutation $\nu_o = (0,1,2,\ldots,n)$ of γ_{n+1}. We have $\varepsilon(\nu_o) = (-1)^n$.
Iterating ν_o we get $\varepsilon(\nu_o^i) = (-1)^{in}$, and in particular

$$(1;a_i,\ldots,a_n,a_o,\ldots,a_{i-1}) = (1;a_{\nu_o^{-i}(0)},\ldots,a_{\nu_o^{-i}(n)}).$$

Let $C = \langle\nu_o\rangle$ be the cyclic subgroup of γ_n generated by ν_o. We are now able to write

$$B(a_o;a_1,\ldots,a_n) = \sum_{\nu\in C} \varepsilon(\nu)(1;a_{\nu^{-1}(0)},\ldots,a_{\nu^{-1}(n)}).$$

With $\gamma_n = \{\sigma \in \gamma_{n+1}: \sigma(0) = 0\}$ we obtain the formula

$$B\gamma(\omega_n) = \sum_{\sigma\in\gamma_n} \varepsilon(\sigma) \sum_{\nu\in C} \varepsilon(\nu)(1;a_{\sigma^{-1}\nu^{-1}(0)},\ldots,a_{\sigma^{-1}\nu^{-1}(n)}) \bmod \bar{B}_{n+1}(A).$$

Since $\gamma_{n+1} = C\gamma_n$, this formula simplifies to

$$B\gamma(\omega_n) = \sum_{\tau\in\gamma_{n+1}} \varepsilon(\tau)(1;a_{\tau^{-1}(0)},\ldots,a_{\tau^{-1}(n)}) \bmod \bar{B}_{n+1}(A)$$

which is precisely what we wanted to show.

<u>Remark 1.2.12</u> The homomorphism $\gamma: \Omega_{A/k} \to H_*(A)$ is an isomorphism if and only if $H_*(A)$ is an exterior algebra over the A-module $H_1(A)$. Assume now A to be a <u>k-flat</u> commutative algebra.
Then $H_*(A)$ equals $\mathrm{Tor}_*^{A\otimes A}(A,A)$, and we have to discuss necessary and sufficient conditions which guarantee that $\mathrm{Tor}_*^{A\otimes A}(A,A)$ is an exterior algebra over $\mathrm{Tor}_1^{A\otimes A}(A,A)$ (note that the multiplicative structure on $\mathrm{Tor}_*^{A\otimes A}(A,A)$ has a resolution-independent definition: we are dealing with the $\pitchfork$-product as defined and discussed in [C.E., ch.XI.; in particular p.217, p.219].).
The most general result we dispose of is the following:

<u>Theorem</u>. Let A be a noetherian k-flat commutative algebra. Then the following two conditions are equivalent:

(1) The A-algebra $H_*(A) = \mathrm{Tor}_*^{A\otimes A}(A,A)$ is the exterior algebra of $\mathrm{Tor}_1^{A\otimes A}(A,A)$, and $\mathrm{Tor}_1^{A\otimes A}(A,A)$ is A-flat

(2) For every prime ideal P of A we have: The local ring A_P is formally smooth over k (see [Ma,ch.11] for a detailed discussion of formal smoothness).

<u>Proof</u>. Clearly, a detailed proof of this equivalence is beyond the scope of these lectures. But it is easy to indicate how available results patch together: By [An,B, 20.31, p.283] our condition (1) is equivalent to (1'): $H_n(A\otimes A,A,-) = 0$ for $n \neq 1$ (where H_n means now n-th simplicial homology functor).

Now, using [An,L, 19.4, p.79], we get: (1') is equivalent to (2'): $H_n(k,A,-) = 0$ for $n \geq 1$. But, again by [An,B, Supp.30,p.331], we get finally: (2') is equivalent to (2).

Assume now k to be a perfect field. Then (2) is equivalent to A being a regular ring (cf. [Ma, 28.M, p.207]). In particular, let A be of finitely generated type over k (i.e. a localization of a homomorphic image of some polynomial ring $R = k[T_1,\ldots,T_n]$ in a finite number of indeterminates over k). If A is k-smooth (i.e. $H_1(A) \simeq \Omega^1_{A/k}$ is locally free), then $H_*(A)$ is an exterior algebra of $H_1(A)$.

We shall come back to this theme in a moment.

<u>Lemma 1.2.13</u> Assume $\mathbb{Q} \subset k$. There is a homomorphism of mixed complexes

$$\mu\colon \overline{C}(A) = (A\otimes\overline{A}^*,b,B) \to \Omega_{A/k} = (\Omega^*_{A/k},0,d)$$

which is given in degree n by the formula

$$\mu(a_o;a_1,\ldots,a_n) = \frac{1}{n!}\, a_o da_1 \wedge\ldots\wedge da_n$$

<u>Proof</u>. First, μ is well-defined, since it is clearly well-defined as a map from A^{*+1} to $\Omega^*_{A/k}$, and since for this map μ we have $D_* \subset \mathrm{Ker}\,\mu$.

We have to show

(1) $\mu\circ b = 0$

(2) $\mu\circ B = d\circ\mu$.

As to (1), recall the operator $\beta : \Omega_* A \to \Omega_* A$ (1.1.11) and the homomorphism of d.g. (cochain) algebras $\pi : \Omega_* A \to \Omega_{A/k}$ (1.1.8).
In degree n we get

$$\mu \circ b(a_o;a_1,\ldots,a_n) = \frac{1}{(n-1)!}\pi\beta(a_o da_1\ldots da_n) = 0$$

since $\beta\Omega_* A \subset [\Omega_* A, \Omega_* A] \subset \operatorname{Ker}\pi$.

As to (2), we get immediately

$$\begin{aligned}(\mu\circ B)(a_o;a_1,\ldots,a_n) &= \frac{1}{(n+1)!}\sum_{i=0}^{n}(-1)^{ni} da_i \wedge\ldots\wedge da_n \wedge da_o \wedge\ldots\wedge da_{i-1} \\ &= \frac{1}{n!}\, da_o \wedge da_1 \wedge\ldots\wedge da_n \\ &= d\left(\frac{1}{n!}\, a_o da_1 \wedge\ldots\wedge da_n\right) \\ &= (d\circ\mu)(a_o;a_1,\ldots,a_n).\end{aligned}$$

Complement 1.2.14 For $\mathbb{Q} \subset k$ we obtain a k-split epimorphism of cochain complexes

$$\mu : (H_*(A), B) \to (\Omega^*_{A/k}, d).$$

Proof. By the skew-commutativity of $\Omega_{A/k}$ we have $\mu\circ\gamma = id_{\Omega_{A/k}}$, as needed.

Definition 1.2.15 k an arbitrary commutative ring, A a unital commutative k-algebra.
The de Rham cohomology $H^*_{DR}(A)$ of A is the cohomology $H^*(\Omega_{A/k}, d)$ of the Kähler-de Rham complex $\Omega_{A/k}$.

Proposition 1.2.16 Assume $\mathbb{Q} \subset k$. The homomorphism of mixed complexes $\mu : \overline{C}(A) \to \Omega_{A/k}$ (cf. 1.2.13) gives rise to a homomorphism of long exact sequences

$$
\begin{array}{ccccccc}
\to H_n(A) & \xrightarrow{I} & HC_n(A) & \xrightarrow{S} & HC_{n-2}(A) & \xrightarrow{B} & H_{n-1}(A) \to \\
\downarrow \mu & & \downarrow \mu & & \downarrow \mu & & \downarrow \mu \\
\to \Omega^n_{A/k} & \longrightarrow & \Omega^n_{A/k}/d\Omega^{n-1}_{A/k} & & & & \Omega^{n-1}_{A/k} \to \\
 & & \oplus & & & \nearrow d & \\
 & & H^{n-2}_{DR}(A) & \hookrightarrow & \Omega^{n-2}_{A/k}/d\Omega^{n-3}_{A/k} & & \\
 & & \oplus & & \oplus & & \\
 & & H^{n-4}_{DR}(A) & = & H^{n-4}_{DR}(A) & & \\
 & & \oplus & & \oplus & & \\
 & & \vdots & & \vdots & &
\end{array}
$$

Proof. The assertion follows from I.2.3.11, provided that we have verified the almost trivial fact, that for the mixed complex $\Omega_{A/k} = (\Omega^*_{A/k}, 0, d)$ homology and cyclic homology are given by

$$H_n(\Omega_{A/k}) = \Omega^n_{A/k}, \quad n \geq 0$$

$$HC_n(\Omega_{A/k}) = \Omega^n_{A/k}/d\Omega^{n-1}_{A/k} \oplus H^{n-2}_{DR}(A) \oplus H^{n-4}_{DR}(A) \oplus \ldots, \quad n \geq 0.$$

The first assertion concerning homology is clear, since our chain differential is zero.
The second assertion concerning cyclic homology is also evident: Look at the associated chain complex

$$
\begin{array}{ccccccccc}
(d\Omega_{A/k})_n & = & \Omega^n_{A/k} & \oplus & \Omega^{n-2}_{A/k} & \oplus & \Omega^{n-4}_{A/k} & \oplus & \ldots \\
d \downarrow & & 0 \downarrow & \swarrow d & 0 \downarrow & \swarrow d & 0 \downarrow & & \\
(d\Omega_{A/k})_{n-1} & = & \Omega^{n-1}_{A/k} & \oplus & \Omega^{n-3}_{A/k} & \oplus & \Omega^{n-5}_{A/k} & \oplus & \ldots
\end{array}
$$

and conclude.

Corollary 1.2.17 In the situation 1.2.16 the following assertions are equivalent:

(1) $\mu: HC_n(A) \to \Omega^n_{A/k}/d\Omega^{n-1}_{A/k} \oplus H^{n-2}_{DR}(A) \oplus H^{n-4}_{DR}(A) \oplus \ldots$

is an isomorphism for every $n \geq 0$.

(2) $\mu : H_n(A) \to \Omega^n_{A/k}$ is an isomorphism for all $n \geq 0$.

(3) $H_*(A)$ is an exterior algebra over $H_1(A)$.

Proof. The equivalence of (1) and (2) follows from I.2.3.15. The equivalence of (2) and (3) is immediate (look at the proof of 1.2.14 and recall 1.2.12).

Theorem 1.2.18 Assume $\mathbb{Q} \subset k$, and let A be a commutative noetherian k-flat algebra such that for every prime ideal P of A the local ring A_P is formally smooth over k. Then

$$HC_n(A) \simeq \Omega^n_{A/k}/d\Omega^{n-1}_{A/k} \oplus H^{n-2}_{DR} \oplus H^{n-4}_{DR} \oplus \ldots$$

for all $n \geq 0$.

Proof. Put 1.2.17 and 1.2.12 together.

Remark 1.2.19 The theorem applies in particular in the following situation: Let k be a field of characteristic 0, and let A be a unital commutative k-algebra of finitely generated type (i.e. a localization of a homomorphic image of some polynomial ring in a finite number of variables over k) which is k-smooth ($\Omega^1_{A/k}$ is locally free). Then, for every $n \geq 0$, $HC_n(A)$ decomposes relative to de Rham cohomology of A, as stated above.

Assume furthermore that $\Omega^1_{A/k}$ is actually A-free of rank m, say. Then $\Omega^n_{A/k} = 0$ for $n > m$. Hence cyclic homology of A stabilizes in the following sense: With $H^{even}_{DR}(A) = \bigoplus_{K \text{ even}} H^K_{DR}(A)$ and $H^{odd}_{DR}(A) = \bigoplus_{K \text{ odd}} H^K_{DR}(A)$ we obtain

$$HC_n(A) = \left\{ \begin{matrix} H^{even}_{DR}(A) & n \text{ even} \\ H^{odd}_{DR}(A) & n \text{ odd} \end{matrix} \right\} \quad n \geq m$$

Examples 1.2.20

(1) $A = \mathbb{C}[T_1,\ldots,T_m]$, the polynomial ring in m variables over $\mathbb{C}$.

$$H^n_{DR}(A) = \begin{cases} \mathbb{C} & n = 0 \\ 0 & n \geq 1 \end{cases} \qquad \text{(Poincaré lemma)}$$

Hence

$$HC_n(A) = \left.\begin{cases} \mathbb{C} & n \text{ even} \\ 0 & n \text{ odd} \end{cases}\right\} \quad n \geq m$$

(For $0 \leq n < m$, $HC_n(A)$ is infinite-dimensional ...)

(2) $A^{(m)} = \mathbb{C}[T_1,\ldots,T_m, T_1^{-1},\ldots,T_m^{-1}]$, the ring of Laurent polynomials in m variables over $\mathbb{C}$.

For $A^{(1)} = \mathbb{C}[T,T^{-1}]$ one sees easily that

$$H^n_{DR}(A^{(1)}) = \begin{cases} \mathbb{C} & n = 0,1 \\ 0 & n \geq 2 \end{cases}$$

I claim that

$$H^n_{DR}(A^{(m)}) = \begin{cases} \mathbb{C}^{\binom{m}{n}} & 0 \leq n \leq m \\ 0 & n \geq m+1 \end{cases}$$

This follows by induction on m using the Künneth-formula

$$H^n_{DR}(A^{(m)}) = \bigoplus_{p+q=n} H^p_{DR}(A^{(m-1)}) \otimes H^q_{DR}(A^{(1)}).$$

We get finally

$$HC_n(A^{(m)}) = \mathbb{C}^{2^{m-1}} \quad \text{for } n \geq m$$

(here, again $HC_n(A^{(m)})$ is infinite-dimensional for $n < m$).

Note that by virtue of I.2.2.2 our results hold for any field k of characteristic 0.

II.2 Relation to Lie theory.

Throughout this paragraph k will be a field of characteristic zero.

II.2.1 Preliminaries around invariant theory.

First we have to recall some basic facts of classical invariant theory.

Remark 2.1.1

(a) For $\Gamma = GL(r,k)$ let $A = k[c_{\mu\nu}: 1 \leq \mu,\nu \leq r]$ be the k-algebra generated by the r^2 coordinate functions on Γ:

$$c_{\mu\nu}(g) = g_{\mu\nu}, \quad 1 \leq \mu,\ \nu \leq r,\ g \in \Gamma.$$

A is actually a bialgebra, with comultiplication $\Delta: A \to A \otimes A$ given by the formula $\Delta(c_{\mu\nu}) = \sum_{\lambda=1}^{r} c_{\mu\lambda} \otimes c_{\lambda\nu}$, and co-unit $\varepsilon : A \to k$ given by $\varepsilon(c_{\mu\nu}) = \delta_{\mu\nu}$ (Δ and ε are k-algebra homomorphisms). A is graded (as a polynomial ring in r^2 variables):

$$A = \bigoplus_{n \geq 0} A(n)$$

Where $A(n)$ is the k-subspace of all polynomial functions in $c_{\mu\nu}$, $1 \leq \mu,\ \nu \leq r$, which are homogeneous of degree n. We have: $\dim_k A(n) = \binom{r^2+n-1}{n}$.

Notation: $[\underline{r},\underline{n}] = \{(i_1,\ldots,i_n): 1 \leq i_j \leq r,\ 1 \leq j \leq n\}$

$i \in [\underline{r},\underline{n}]$ is a multi-index of length $n \geq 1$.

$$c_{ij} := c_{i_1 j_1} c_{i_2 j_2} \ldots c_{i_n j_n}, \quad i,j \in [\underline{r},\underline{n}]$$

The multiplicativity of Δ and ε implies

$$\Delta(c_{ij}) = \sum_{s \in [\underline{r},\underline{n}]} c_{is} \otimes c_{sj}, \quad \varepsilon(c_{ij}) = \delta_{ij}$$

for $i,j \in [\underline{r},\underline{n}]$ (where $\delta_{ij} = \begin{cases} 1 & i = j \\ 0 & i \neq j \end{cases}$ as n-tuples)

Thus $A(n)$ is a finite-dimensional subcoalgebra of A.

(b) Consider now its dual $S(n) = \mathrm{Hom}_k(A(n),k)$ with the convolution product

$\xi * \eta = (\xi \otimes \eta) \circ \Delta$; more explicitely

$$(\xi * \eta)(c) = \sum_t \xi(c_t)\eta(c_t'), \text{ where } \Delta(c) = \sum_t c_t \otimes c_t'.$$

$S(n)$ is a finite-dimensional associative k-algebra with unit element ε, the co-unit of $A(n)$. $S(n)$ is called the Schur algebra (of order n) (relative to $\Gamma = GL(r,k)$)

Let $k[\Gamma]$ be the group algebra of $\Gamma = GL(r,k)$. There is an evaluation homomorphism

$$e: k[\Gamma] \to S(n)$$

given by $e(g)(c) = c(g)$ for $c \in A(n)$, $g \in \Gamma$ (and extended by linearity).

(c) Proposition: $e: k[\Gamma] \to S(n)$ is surjective. Moreover, for $f \in A = k[c_{\mu\nu} : 1 \le \mu,\nu \le r]$ the following two conditions are equivalent :

(i) $f \in A(n)$

(ii) f vanishes on $\mathrm{Ker}\ e$

Proof. [Gr, p.24]

Complement: Let W be a left $k[\Gamma]$-module. Then the following two conditions are equivalent:

(i) $\mathrm{Ker}\ e.W = 0$ $(e: k[\Gamma] \to S(n))$

(ii) All representative functions relative to

$k[\Gamma] \to \mathrm{End}_k W$ lie in $A(n)$.

Proof. [Gr, p.24]

As a consequence, the categories of left $S(n)$-modules and of left $k[\Gamma]$-modules having representative functions in $A(n)$ are equivalent.

Remark 2.1.2 A first visit of γ_n, the symmetric group of order n.

Write $I = [\underline{r},\underline{n}] = \{(i_1,\ldots,i_n): 1 \le i_j \le r,\ 1 \le j \le n\}$

γ_n operates on the right via place-permutations on I:

$$i\sigma = (i_{\sigma(1)},\dots,i_{\sigma(n)}) \quad \text{for} \quad \sigma \in \gamma_n,\ i = (i_1,\dots,i_n) \in I$$

γ_n operates equally on the right on $I \times I$:

$$(i,j)\sigma = (i\sigma, j\sigma)$$

$i \sim j$ or $(i,j) \sim (K,\ell)$ means orbital equivalence, as usual.

Example:
Consider the generating functions c_{ij}, $(i,j) \in I \times I$ of $A(n)$:

$$c_{ij} = c_{i_1j_1}c_{i_2}c_{j_2}\dots c_{i_nj_n}$$

$$c_{ij} = c_{K\ell} \quad \text{if and only if} \quad (i,j) \sim (K,\ell)$$

As a consequence, the number of γ_n-orbits of $I \times I$ is equal to $\binom{r^2+n-1}{2}$.

Note that our identification rule passes to the dual $S(n)$: $S(n)$ has a k-basis $\{\xi_{ij}: (i,j) \in I \times I\}$ dual to $\{c_{ij}:(i,j) \in I \times I\}$ of $A(n)$.
Here again: $\xi_{ij} = \xi_{K\ell} \Leftrightarrow (i,j) \sim (K,\ell)$

<u>Remark 2.1.3</u> The $k[\Gamma]$-$k[\gamma_n]$ bimodule $V^{\otimes n}$, $n \geq 1$. $\Gamma = GL(r,k)$, $V = ke_1 \oplus \dots \oplus ke_r$ an r-dimensional k-vector space.
Γ acts on V:

$$ge_\nu = \sum_\mu g_{\mu\nu}e_\mu = \sum_\mu c_{\mu\nu}(g)e_\mu$$

This gives a natural action of Γ on $V^{\otimes n}$, $n \geq 1$:
Consider the k-basis $\{e_i, i \in I = [\underline{r},\underline{n}]\}$ of $V^{\otimes n}$, induced by the initial k-basis of V:

$$e_i = e_{i_1} \otimes \dots \otimes e_{i_n} \quad \text{for} \quad i = (i_1,\dots i_n)$$

$$ge_j = ge_{j_1} \otimes \dots \otimes ge_{j_n} = \sum_{i \in I} g_{i_1j_1}\dots g_{i_nj_n}e_i = \sum_{i \in I} c_{ij}(g)e_i$$

All representative functions on $W = V^{\otimes n}$ are thus elements of $A(n)$;

this means that $V^{\otimes n}$ is actually an $S(n)$-module:

$$\xi e_j = \sum_{i \in I} \xi(c_{ij}) e_i \quad \text{for all} \quad \xi \in S(n)$$

Note that $k[\Gamma]$ and $S(n)$ act by the same k-algebra of linear transformations on $W = V^{\otimes n}$.
On the other hand, the symmetric group γ_n acts on (the right of) $V^{\otimes n}$ by means of place-permutations:

$$e_i \sigma = e_{i\sigma}, \quad i \in I = [\underline{r},\underline{n}], \quad \sigma \in \gamma_n$$

More explicitely:

$$(e_{i_1} \otimes \ldots \otimes e_{i_n})\sigma = e_{i_{\sigma(1)}} \otimes \ldots \otimes e_{i_{\sigma(n)}}$$

Since $(\xi x)\sigma = \xi(x\sigma)$ for all $\xi \in S(n)$, $x \in W = V^{\otimes n}$ and $\sigma \in \gamma_n$

$V^{\otimes n}$ becomes an $S(n)$-$k[\gamma_n]$ bimodule (equivalently: a $k[\Gamma]$-$k[\gamma_n]$ bimodule, such that the representative functions relative to the action of Γ are all elements of $A(n)$).

<u>Proposition 2.1.4</u>

(a) The representation of $S(n)$ as a k-algebra of linear transformations on $V^{\otimes n}$ is faithful, more precisely: $S(n) \simeq \mathrm{End}_{k[\gamma_n]} V^{\otimes n}$

(b) On the other hand, the image of $k[\gamma_n]$ in $\mathrm{End}_k V^{\otimes n}$ is $\mathrm{End}_{S(n)} V^{\otimes n}$ $= \mathrm{End}_{k[\Gamma]} V^{\otimes n}$.

<u>Proof</u>.

(a) cf. [Gr,(2.6c),p.28/29]

(b) This is an immediate consequence of the density theorem (cf. [Be, p.89]): $W = V^{\otimes n}$ is a completely reducible $k[\gamma_n]$-module, and $S(n)$ is its centralizer. Hence the image of $k[\gamma_n]$ in $\mathrm{End}_k V^{\otimes n}$ is dense in the double centralizer $\mathrm{End}_{S(n)} V^{\otimes n}$, which gives our assertion, since we are dealing with finite-dimensional vector spaces (and algebras), where density means equality.

<u>Complement 2.1.5</u> Suppose now $r \geq n$ ($r = \dim_k V$).

Then $k[\gamma_n] \to \mathrm{End}_k V^{\otimes n}$ is injective, and thus

$$k[\gamma_n] \simeq \mathrm{End}_{k[\Gamma]} V^{\otimes n}$$

<u>Proof</u>. Write $\underline{\sigma}$ for the k-automorphism induced by $\sigma \in \gamma_n$ on $V^{\otimes n}$. Consider n k-linear independent vectors $e_1,\ldots,e_n$ of V, and put $e_* = e_1 \otimes\ldots\otimes e_n$. Then $\{e_*\sigma, \sigma \in \gamma_n\}$ is a set of k-linear independent elements of $V^{\otimes n}$. A relation $\Sigma\alpha_\sigma\underline{\sigma} = 0$ gives rise (evaluate on e_*) to a relation $\Sigma\alpha_\sigma e_*\sigma = 0$, and we are through.

<u>Remark 2.1.6</u> Associate with $V = ke_1 \oplus\ldots\oplus ke_r$ its Lie algebra of endomorphisms $g = (\mathrm{End}_k V, [,])$. Consider $E_{\mu\nu} \in g$, defined by $E_{\mu\nu}e_\lambda = \delta_{\nu\lambda}e_\mu$ for $1 \le \lambda,\mu,\nu \le r$.

Identify now $g^n = g^{\otimes n}$ with $\mathrm{End}_k V^{\otimes n}$ in the usual way:

A k-basis of $\mathrm{End}_k V^{\otimes n}$ is given by

$$\{E_{ij}: i,j \in I = [\underline{r},\underline{n}]\}$$

where $E_{ij} = E_{i_1j_1} \otimes\ldots\otimes E_{i_nj_n}$ for $i = (i_1,\ldots,i_n)$, $j = (j_1,\ldots,j_n)$

hence $E_{ij}e_K = \delta_{jK}e_i$ for all multi-indices $i,j,K \in I = [\underline{r},\underline{n}]$

It is clear that for $\sigma \in \gamma_n$ its image $\underline{\sigma} \in g^{\otimes n}$ is given by the formula

$$\underline{\sigma} = \sum_{i\in I} E_{i\sigma,i}$$

$\underline{\sigma}$ acts as a place-permutation on $V^{\otimes n}$.

Assume now for greater ease that $r \ge n$, i.e. that, without loss of generality, $k[\gamma_n] \subset g^n$.

Let us look at the effect of conjugation by $\pi \in \gamma_n$, from the right:

$$\underline{\pi\sigma\pi}^{-1} = \sum_{i\in I} E_{i\pi^{-1}\sigma\pi,i} = \sum_{j\in I} E_{j\sigma\pi,j\pi} = \sum_{j\in I} E_{(j\sigma,j)\pi}$$

This means that conjugation with π (from the right) in $k[\gamma_n]$ translates to place-permutation (according to π) relative to the embedding

$$k[\gamma_n] \subset g^n.$$

<u>Caution</u>: Every $\sigma \in \gamma_n$ acts as a place-permutation (on $V^{\otimes n}$), but

may be place-permuted itself (as an element of g^n).

Proposition 2.1.7 In the situation 2.1.6, let g act on $g^n = g^{\otimes n}$ via the adjoint representation:

$$X.(T_1 \otimes T_2 \otimes \ldots \otimes T_n) = [X,T_1] \otimes T_2 \otimes \ldots \otimes T_n + T_1 \otimes [X,T_2] \otimes \ldots \otimes T_n + \ldots + T_1 \otimes T_2 \otimes \ldots \otimes [X,T_n]$$

Then $(g^n)^g$, the submodule of invariant elements of g^n, identifies with the image of $k[\gamma_n]$ as k-algebra of linear transformations on $W = V^{\otimes n}$. If $r = \dim_k V \geq n$, then $k[\gamma_n] \simeq (g^n)^g$.

Proof.

(1) g identifies with $L(\Gamma)$, the Lie algebra of the algebraic k-group $\Gamma = GL(r,k)$. We have the adjoint representation

$$\mathrm{Ad}: \Gamma \to GL(g)$$

$$g \to [T \to gTg^{-1}]$$

with differential

$$\mathrm{ad}: g \to \mathrm{End}_k(g)$$

$$X \to [T \to [X,T] = XT - TX].$$

Consider now $g^n = g^{\otimes n}$ with the adjoint representations extended on the tensor product:

$$g.(T_1 \otimes T_2 \otimes \ldots \otimes T_n) = gT_1g^{-1} \otimes gT_2g^{-1} \otimes \cdots \otimes gT_ng^{-1}$$

$$X.(T_1 \otimes T_2 \otimes \ldots \otimes T_n) = [X,T_1] \otimes T_2 \otimes \cdots \otimes T_n + \ldots + T_1 \otimes T_2 \otimes \ldots \otimes [X,T_n]$$

$(g^n)^\Gamma$, the space of invariant elements under the action of Γ, equals $(g^n)^g$, the space of invariant elements under the action of g (cf. [Hu, p.88]):

$$\{Z \in g^n : g.Z = Z \text{ for all } g \in \Gamma\} = \{Z \in g^n : X.Z = 0 \text{ for all } X \in g\}$$

(2) Our initial representation

$k[\Gamma] \to \mathrm{End}_k V^{\otimes n} = g^n$ is given by

$g \to \hat{g} = g \otimes g \otimes \ldots \otimes g$ (n times)

Hence we get for the adjoint representation of Γ on g^n:

$$g.(T_1 \otimes \ldots \otimes T_n) = \hat{g}(T_1 \otimes \ldots \otimes T_n)\hat{g}^{-1}$$

This means that

$$(g^n)^g = (g^n)^\Gamma = \{Z \in \mathrm{End}_k\ V^{\otimes n} : \hat{g}Z\hat{g}^{-1} = Z \text{ for all } g \in \Gamma\}$$

i.e.

$$(g^n)^g = \{Z \in \mathrm{End}_k V^{\otimes n} : \hat{g}Z = Z\hat{g} \text{ for all } g \in \Gamma\}$$

$$= \mathrm{End}_{k[\Gamma]} V^{\otimes n}$$

$$= \mathrm{Im}(k[\gamma_n] \to g^n)$$

by virtue of 2.1.4.
This finishes the proof of our proposition.

Remark 2.1.8

(1) Let g be any Lie algebra over k, and let $U(g)$ be its universal enveloping algebra.
The left (right) $U(g)$-modules are precisely the left (right) g-modules (equivalently: the antisymmetric g-g bimodules).
The augmentation homomorphism $U(g) \to k$ corresponds to the trivial g-module structure on k (cf. [C.E., pp.266-274] for details).
For every g-module (equivalently: left (right) $U(g)$-module) W we get the following identifications

$W^g = \mathrm{Hom}_{U(g)}(k,W)$, the submodule of invariants of W

$W_g = W \otimes_{U(g)} k = W/Wg$, the factor module of coinvariants of W

Combining with k-duals, we obtain

$$(W^*)^g = \mathrm{Hom}_{U(g)}(k, \mathrm{Hom}_k(W,k))$$

$$= \mathrm{Hom}_k(W \underset{U(g)}{\otimes} k,\ k)$$

$$= (W_g)^*$$

Note that g acts on $\mathrm{Hom}_k(W,k)$ by the formula

$$(X.\lambda)(w) = -\lambda(X.w)$$

(2) For $r = \dim_k V \geq n$, $g = (\mathrm{End}_k V, [\ ,\ ])$
we have, according to 2.1.7, the isomorphism

$k[\gamma_n] \overset{\sim}{\to} (g^n)^g$, and, by duality,

$$[(g^n)^g]^* \overset{\sim}{\to} k[\gamma_n]^*$$

We identify g with g^*, and $k[\gamma_n]$ with $k[\gamma_n]^*$ (more explicitely: in terms of dual bases, we identify $E^*_{\mu\nu} \in g^*$ with $E_{\nu\mu} \in g$, and $\sigma^* \in k[\gamma_n]^*$ with $\sigma \in k[\gamma_n]$).
This gives finally, by virtue of (1), an isomorphism

$$(g^n)_g \simeq k[\gamma_n]$$

which identifies σ with $\sum_{i \in I} E_{i,i\sigma}$ (cf. 2.1.6)
(we shall suppress equivalence class notation).
Let us point out, once more, that action of $\pi \in \gamma_n$ by conjugation from the right on $k[\gamma_n]$ corresponds to place-permutation according to π on g^n.

(3) For $r = \dim_k V \geq n$, the isomorphism

$$k[\gamma_n] \overset{\sim}{\to} (g^n)_g$$

"absorbs" the size of g.
Let us make this statement more precise.
Consider $g_r = gl_r(k) = (M_r(k), [\ ,\ ])$, $r \geq n$.
The standard inclusions $g_r \subset g_{r+1}$ define $g = g_\infty = gl(k) = \underset{\to}{\lim}\, g_r$,

the Lie algebra of infinite matrices with only a finite number of non-zero entries. Since universal envelopes commute with direct limits, we have $U(g) = \varinjlim U(g_r)$, and finally

$$(g^n)_g = (\varinjlim g_r)^n \underset{\varinjlim U(g_r)}{\otimes} k = \varinjlim (g_r^n)_{g_r}$$

The isomorphisms $k[\gamma_n] \overset{\sim}{\to} (g_r^n)_{g_r}$, $r \geq n$ are obviously compatible with the inclusions $g_r \subset g_{r+1}$, i.e. we get a commutative diagram of isomorphisms

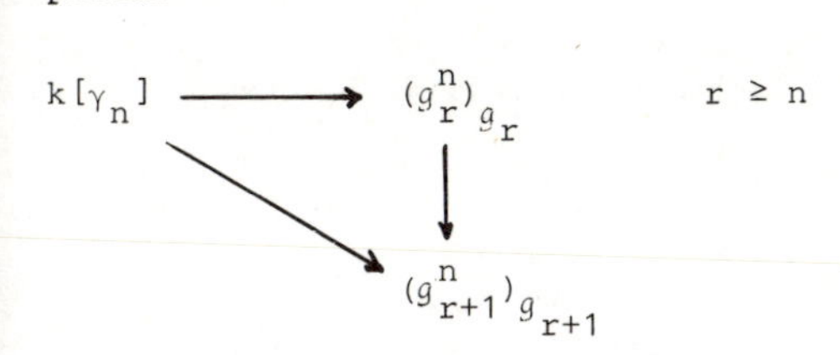

(Note that $\sigma \in k[\gamma_n]$ acts as a place-permutation $\underline{\sigma}_r$ on $V_r^{\otimes n}$, where V_r is the k-vector space of $r\times 1$-matrices (columns) with coefficients in k; hence $\underline{\sigma}_r = \underline{\sigma}_{r+1}/V_r^{\otimes n}$, relative to the inclusion $V_r^{\otimes n} \subset V_{r+1}^{\otimes n}$)
In particular, fixing the initial isomorphism $(r = n)$ and passing to the limit, we get a commutative diagram of isomorphisms

$$\begin{array}{ccc} k[\gamma_n] & \longrightarrow & (g_n^n)_{g_n} \\ & \searrow & \downarrow \\ & & (g_\infty^n)_{g_\infty} \end{array}$$

This means (in matrix notation) that the upper horizontal isomorphism

$$\sigma \to \sum_{i\in[\underline{n},\underline{n}]} E_{i,i\sigma}$$

already defines the limit isomorphism $k[\gamma_n] \simeq (g^n)_g$; i.e., when dealing with coinvariants $(g^n)_g$ for $g = \varinjlim gl_r(k)$, the left upper $n\times n$ square $gl_n(k) \subset g$ parametrizes all of $(g^n)_g$.

Remark-Definition 2.1.9 g a Lie algebra over k, $U(g)$ its universal enveloping algebra (with augmentation $U(g) \to k$)

Define, for every right $\mathfrak{g}$-module W, the <u>homology of $\mathfrak{g}$ with coefficients in W</u> as $H_*(\mathfrak{g},W) = \mathrm{Tor}_*^{U(\mathfrak{g})}(W,k)$. (cf. [C.E., p.270]).
In particular: $H_*(\mathfrak{g}) = \mathrm{Tor}_*^{U(\mathfrak{g})}(k,k)$ is the <u>homology of $\mathfrak{g}$</u> (with coefficients in k).
Recall ([C.E.,p.282] or [H.St.,p.239]): $H_n(\mathfrak{g})$ is the n-th homology group of the complex $(E(\mathfrak{g}),d)$, where

(i) $E_n(\mathfrak{g}) = \Lambda^n\mathfrak{g}$ is the n-th exterior power of $\mathfrak{g}$ over k, $n \geq 0$.

(ii) the differential d is given by the following formula:

$$d(x_1 \wedge \ldots \wedge x_n) = \sum_{1 \leq i < j \leq n} (-1)^{i+j}[x_i,x_j] \wedge x_1 \wedge \ldots \wedge \hat{x}_i \wedge \ldots \wedge \hat{x}_j \wedge \ldots \wedge x_n$$

(conventionally $\hat{x}_i$ means "omit x_i").

<u>Properties 2.1.10</u> (of $(E(\mathfrak{g}),d)$)

(1) Let, for every $x \in \mathfrak{g}$, $\theta(x)$ be the k-derivation of the exterior algebra $E(\mathfrak{g})$ which prolongs the inner derivation $\mathrm{ad}x$ on $\mathfrak{g}$.
We have explicitely:

$$\theta(x)(x_1 \wedge \ldots \wedge x_n) = [x,x_1] \wedge x_2 \wedge \ldots \wedge x_n + x_1 \wedge [x,x_2] \wedge \ldots \wedge x_n + \ldots + x_1 \wedge \ldots \wedge [x,x_n]$$

Note the equality:

$$\theta([x,y]) = \theta(x)\theta(y) - \theta(y)\theta(x) \quad \text{for all } x,y \in \mathfrak{g}.$$

(cf. [H.St., p.240]).

$\theta: \mathfrak{g} \to \mathrm{Der}_k(E(\mathfrak{g}))$ defines a representation of $\mathfrak{g}$ as a Lie algebra of k-derivations on $E(\mathfrak{g})$. In particular, for every subalgebra $\mathfrak{h}$ of $\mathfrak{g}$, $E(G)$ is a (left) $\mathfrak{h}$-module via the induced representation

$$\theta: \mathfrak{h} \to \mathrm{Der}_k(E(\mathfrak{g})).$$

(2) For every $u \in E(\mathfrak{g})$ there is an endomorphism $\sigma(u) \in \mathrm{End}_k E(\mathfrak{g})$, defined by $\sigma(u)(v) = u \wedge v$ (left multiplication by u).
The following identity is easily verified:

$$\theta(x) + \sigma(x)d + d\sigma(x) = 0 \quad \text{for all } x \in \mathfrak{g}$$

(cf. [H.St., p.241]).

Note two consequences:

(2.1) $\theta(x)d = d\theta(x)$ for all $x \in g$, i.e. d is a g-homomorphism.

(2.2) With $Z_*(g) = \text{Ker } d$, $B_*(g) = \text{Im } d$, we get $\theta(x)Z_*(g) \subset B_*(g)$ for all $x \in g$, i.e. g operates trivially on $H_*(g) = Z_*(g)/B_*(g)$.

Proposition 2.1.11 g a Lie algebra over k, h a sub-Lie algebra of g. Assume $E_n(g)$ semi-simple as an h-module for all $n \geq 0$. Then, taking coinvariants with respect to h gives a homomorphism of chain complexes

$$(E(g),d) \longrightarrow (E(g)_h,d)$$

which is a quasi-ismorphism.

Proof. h acts on $E(g)$ via $\theta: h \to \text{Der}_k(E(g))$, as noted in 2.1.10(1). Define a subcomplex (L_*,d) of $(E(g),d)$ by $L_n = h.E_n(g)$, $n \geq 0$; the h-linearity of d guarantees that $d(L_n) \subset L_{n-1}$, $n \geq 1$; hence we get an exact sequence of chain complexes

$$0 \longrightarrow (L_*,d) \longrightarrow (E(g),d) \longrightarrow (E(g)_h,d) \longrightarrow 0.$$

We need only show that L_* is acyclic; the long exact homology sequence then yields our assertion.
The semi-simplicity of $E_n(g)$ as an h-module gives rise to an h-direct decomposition

$$E_n(g) = L_n \oplus E_n(g)_h, \quad n \geq 0$$

and $L_n = \bigoplus_i S_{n,i}$ is a direct sum of simple h-modules on which h acts non-trivially.

Since our short exact sequence of chain complexes is h-split, the homology $H_*(g)$ decomposes according to the h-direct decomposition of $E(g)$. But h acts trivially on homology (2.1.10(2.2)), thus the simple factors of L_n on which h acts non-trivially must all disappear when passing to homology. This means that L_* must be acyclic.

Application 2.1.12 Let A be a unital associative k-algebra, and consider $gl(A) = g \otimes A$, where $g = \lim_{\to} gl_r(k)$ is the k-algebra of infinite matrices which have only a finite number of non-zero entries.

Then $(E(gl(A)),d) \to (E(gl(A))_{\mathfrak{g}},d)$

is a quasi-isomorphism.

Proof.

(1) Some preliminaries:

Put $\mathfrak{g}_r = gl_r(k)$, with $1 \le r \le \infty$ $(\mathfrak{g}_\infty = \mathfrak{g} = \varinjlim \mathfrak{g}_r)$, and let (sgn) be the 1-dimensional γ_n-module on which γ_n acts by signature: $\sigma.1 = \varepsilon(\sigma)1$, $\sigma \in \gamma_n$.

Then $E_n(\mathfrak{g}_r \otimes A) \simeq (\mathfrak{g}_r^n \otimes A^n) \underset{\gamma_n}{\otimes} (sgn)$

(where γ_n acts on $\mathfrak{g}_r^n \otimes A^n = \mathfrak{g}_r^{\otimes n} \otimes A^{\otimes n}$ by parallel place-permutation:

$$[(x_1,\ldots,x_n) \otimes (a_1,\ldots,a_n)]\sigma = (x_{\sigma(1)},\ldots,x_{\sigma(n)}) \otimes (a_{\sigma(1)},\ldots, a_{\sigma(n)})$$

This isomorphism is an immediate consequence of the very definition of $\Lambda^n(\mathfrak{g}_r \otimes A)$ as a quotient of $(\mathfrak{g}_r \otimes A)^n = (\mathfrak{g}_r \otimes A)^{\otimes n}$: divide out the k-linear span of all $z_1 \otimes\ldots\otimes z_n - \varepsilon(\sigma)z_{\sigma(1)} \otimes\ldots\otimes z_{\sigma(n)}$, $\sigma \in \gamma_n$.

Now, comparing the adjoint action $\theta(x)$ of $x \in \mathfrak{g}_r$ on $E_n(\mathfrak{g}_r \otimes A)$ and the adjoint action adx of $x \in \mathfrak{g}_r$ on $\mathfrak{g}_r^n = \mathfrak{g}_r^{\otimes n}$, we get, by transport on the right side of our isomorphism:

$$\theta(x).[(x_1,\ldots,x_n) \otimes (a_1,\ldots,a_n) \otimes 1] = [adx.(x_1,\ldots,x_n)] \otimes (a_1,\ldots,a_n) \otimes 1$$

Consequences:

(i) $E_n(\mathfrak{g}_r \otimes A)$ is semi-simple under $\theta: \mathfrak{g}_r \to End_k(E_n(\mathfrak{g}_r \otimes A))$ provided $\mathfrak{g}_r^n = \mathfrak{g}_r^{\otimes n}$ is semi-simple under ad: $\mathfrak{g}_r \to End_k(\mathfrak{g}_r^n)$

(ii) $E_n(\mathfrak{g}_r \otimes A)_{\mathfrak{g}_r} \simeq [(\mathfrak{g}_r^n)_{\mathfrak{g}_r} \otimes A^n] \underset{\gamma_n}{\otimes} (sgn)$

(for the coinvariants)

(2) Let r be finite. We shall first show that

$(E(gl_r(A)),d) \to (E(gl_r(A))_{\mathfrak{g}_r},d)$

is a quasi-ismorphism.

(where $gl_r(A) = g_r \otimes A = (M_r(A),[\,,\,]))$

According to 2.1.11 and (1), consequence (i), we have only to make sure that g_r^n is semi-simple under $ad: g_r \to End_k(g_r^n)$. Note that this is equivalent to saying that g_r^n is semi-simple under $Ad: GL(r,k) \to GL(g_r^n)$ (cf. the arguments in the proof of 2.1.7). Our assertion follows now from the linear reductivity of $GL(r,k)$ (cf. [Fo, p.146]).

(3) We have now to pass to the direct limit.

The commutative squares (of homomorphisms of chain complexes)

$$\begin{array}{ccc} (E(gl_{r+1}(A)),d) & \longrightarrow & (E(gl_{r+1}(A))_{g_{r+1}},d) \\ \uparrow & & \uparrow \\ (E(gl_r(A)),d) & \longrightarrow & (E(gl_r(A))_{g_r},d) \end{array}$$

allow to pass to the direct limits.

By virtue of (1), the direct limit arrow identifies with

$$(E(gl(A)),d) \to (E(gl(A))_g,d).$$

Since homology commutes with direct limits, we get finally our assertion.

II.2.2. Cyclic homology and the Lie algebra homology of matrices.

<u>Situation 2.2.1</u> k a field of characteristic zero. A a unital associative k-algebra.

$M_r(A)$ the k-algebra of $r\times r$ matrices with coefficients in A.

$M(A) = M_\infty(A) = \varinjlim M_r(A)$ the k-algebra of infinite matrics which have only a finite number of non-zero A-entries.

$gl_r(A) = (M_r(A),[\,,\,])$ the Lie algebra of $r\times r$ matrices with coefficients in A.

The standard inclusions $gl_r(A) \subset gl_{r+1}(A)$ define $gl(A) = \varinjlim gl_r(A) = (M(A),[\,,\,])$

Notation:

$\mathfrak{g}_r = gl_r(k)$ for $1 \le r \le \infty$

$\mathfrak{g} = \mathfrak{g}_\infty = gl_\infty(k) = (M(k),[\,,\,])$

Note that

$gl_r(A) = \mathfrak{g}_r \otimes A \quad 1 \le r \le \infty$, in particular

$gl(A) = \mathfrak{g} \otimes A$.

Lemma 2.2.2 Consider, for $1 \le r \le \infty$, the sequence of maps

$$\lambda_n: E_{n+1}(gl_r(A)) \to C_n^\lambda(M_r(A))$$

defined by

$$\lambda_n(x_o \wedge \ldots \wedge x_n) = (-1)^n \sum_{\sigma\in\gamma_n} \varepsilon(\sigma)(x_o, x_{\sigma 1}, \ldots, x_{\sigma(n)}) \bmod (1-t)$$

$$\lambda_*: (E(gl_r(A))[-1], d[-1]) \to C_*^\lambda(M_r(A))$$

is a homomorphism of chain complexes.

Proof. Note first the dimension shift; furthermore, observe that for $r = \infty$ we are in the setting of non-unital cyclic homology.
λ is well-defined, thanks to the cyclic permutation relation

$$(a_o, a_1, \ldots, a_n) = (-1)^n (a_n, a_o, \ldots, a_{n-1}) \bmod (1-t),$$

on the right side. We shall consider our generator $t \in G_{n+1}$ as the cycle $t = (0,1,\ldots,n)$ of length $n+1$ in $\gamma_{n+1} = \gamma_{\{o,1,\ldots,n\}}$ (we need t for the transcription of $(x_n, x_{\sigma o}, \ldots, x_{\sigma(n-1)})$-tensors to $(x_o, x_{\sigma 1}, \ldots, x_{\sigma n})$-tensors; cf. the detailed arguments below).
In the sequel we shall drop the "mod(1-t)"-notation.

Claim. $b\lambda(x_o \wedge \ldots \wedge x_n) = \lambda d(x_o \wedge \ldots \wedge x_n)$

$$= (-1)^n \sum_{\nu\in\gamma_n} \varepsilon(\nu)(x_{\nu_o} x_{\nu_1}, x_{\nu_2}, \ldots, x_{\nu_n})$$

in $C_{n-1}^\lambda(M_r(A))$.

For $t = (0,1,\ldots,n)$ we have $\varepsilon(t^K) = (-1)^{Kn}$.

Decompose $\gamma_{n+1} = \gamma^o_{n+1} \cup \gamma^1_{n+1} \cup \ldots \cup \gamma^n_{n+1}$

where $\gamma^K_{n+1} = \{\nu \in \gamma_{n+1}: \nu(K) = 0\}$, $0 \leq K \leq n$ (Convention: $\gamma^{n+1}_{n+1} = \gamma^o_{n+1}$ $= \gamma_n$).

We get a bijection $\gamma^o_{n+1} \ni \sigma \to \nu = \sigma \circ t^i \in \gamma^{n+1-i}_{n+1}$.

Now, let us write down

$$b\lambda(x_o \wedge \ldots \wedge x_n) = (-1)^n \sum_{\sigma \in \gamma_n} \varepsilon(\sigma) b(x_o, x_{\sigma(1)}, \ldots, x_{\sigma(n)})$$

$$= (-1)^n \sum_{\sigma \in \gamma^o_{n+1}} \varepsilon(\sigma) \Big(\sum_{i=0}^{n-1} (-1)^i (x_{\sigma(o)}, \ldots x_{\sigma(i)} x_{\sigma(i+1)} \ldots, x_{\sigma(n)})$$

$$+ (-1)^n (x_{\sigma(n)} x_{\sigma(o)}, x_{\sigma(1)}, \ldots, x_{\sigma(n-1)}) \Big)$$

We have in $C^\lambda_{n-1}(M_r(A))$:

$$(x_{\sigma(o)}, \ldots, x_{\sigma(i)} x_{\sigma(i+1)}, \ldots, x_{\sigma(n)})$$

$$= (-1)^{i(n-1)} (x_{\sigma(i)} x_{\sigma(i+1)}, \ldots, x_{\sigma(n)}, x_{\sigma(o)}, \ldots, x_{\sigma(i-1)})$$

$$= (-1)^{i(n-1)} (x_{\sigma t^i(o)} x_{\sigma t^i(1)}, \ldots, x_{\sigma t^i(n-i)}, x_o, \ldots, x_{\sigma t^i(n)})$$

Consequently:

$$\sum_{\sigma \in \gamma^o_{n+1}} \varepsilon(\sigma) (-1)^i (x_{\sigma(o)}, \ldots, x_{\sigma(i)} x_{\sigma(i+1)}, \ldots, x_{\sigma(n)})$$

$$= \sum_{\sigma \in \gamma^o_{n+1}} \varepsilon(\sigma) (-1)^{in} (x_{\sigma t^i(o)} x_{\sigma t^i(1)}, \ldots, x_{\sigma t^i(n)})$$

$$= \sum_{\sigma \in \gamma^o_{n+1}} \varepsilon(\sigma t^i) (x_{\sigma t^i(o)} x_{\sigma t^i(1)}, x_{\sigma t^i(2)}, \ldots, x_{\sigma t^i(n)})$$

$$= \sum_{\nu \in \gamma^{n+1-i}_{n+1}} \varepsilon(\nu) (x_{\nu(o)} x_{\nu(1)}, x_{\nu(2)}, \ldots, x_{\nu(n)})$$

Analogously:

$$\sum_{\nu \in \gamma^o_{n+1}} \varepsilon(\sigma) (-1)^n (x_{\sigma(n)} x_{\sigma(o)}, x_{\sigma(1)}, \ldots, x_{\sigma(n-1)})$$

$$= \sum_{\nu\in\gamma^1_{n+1}} \varepsilon(\nu)(x_{\nu(o)}x_{\nu(1)},x_{\nu(2)},\dots,x_{\nu(n)})$$

and finally

$$b\lambda(x_o\wedge\dots\wedge x_n) = (-1)^n \sum_{\nu\in\gamma_{n+1}} \varepsilon(\nu)(x_{\nu(o)}x_{\nu(1)},x_{\nu(2)},\dots,x_{\nu(n)}).$$

On the other hand

$$\lambda d(x_o\wedge\dots\wedge x_n) = (-1)^{n-1} \sum_{0\le i<j\le n} (-1)^{i+j} \sum_{\sigma\in\gamma^{i,j}_{n+1}} \varepsilon(\sigma)([x_i,x_j],x_{\sigma(o)},\dots,x_{\sigma(n)})$$

where $\gamma^{i,j}_{n+1} = \{\sigma \in \gamma_{n+1}: \sigma\{i,j\} = \{i,j\}\}, \quad i < j$

hence:

$$\lambda d(x_o\wedge\dots\wedge x_n) = (-1)^{n-1} \sum_{0\le i<j\le n} (-1)^{i+j} \sum_{\sigma\in\gamma^{i,j}_{n+1}} \varepsilon(\sigma)(x_{\sigma i}\ x_{\sigma j},x_{\sigma o},\dots,x_{\sigma(n)})$$

Define $\gamma^{(i,j)}_{n+1} = \{\nu \in \gamma_{n+1}: \nu\{0,1\} = \{i,j\}\}$, $i < j$

$\gamma_{n+1} = \bigcup_{0\le i<j\le n} \gamma^{(i,j)}_{n+1}$, and $\gamma^{i,j}_{n+1}$ is in bijection with $\gamma^{(i,j)}_{n+1}$ via composition with the permutation θ_{ij} which maps $(0,1,2,\dots,n)$ onto $(i,j,0,\dots\hat{i},\dots\hat{j},\dots n)$. $\varepsilon(\theta_{ij}) = (-1)^{i+j-1}$.

Now we are able to conclude:

$$\lambda d(x_o\wedge\dots\wedge x_n) = (-1)^n \sum_{0\le i<j\le n} \sum_{\sigma\in\gamma^{i,j}_{n+1}} \varepsilon(\sigma)(-1)^{i+j-1}(x_{\sigma\theta_{ij}o}x_{\sigma\theta_{ij}1},\dots x_{\sigma\theta_{ij}n})$$

$$= (-1)^n \sum_{0\le i<j\le n} \sum_{\nu\in\gamma^{(i,j)}_{n+1}} \varepsilon(\nu)(x_{\nu o}x_{\nu 1},x_{\nu 2},\dots,x_{\nu n})$$

$$= (-1)^n \sum_{\nu\in\gamma_{n+1}} \varepsilon(\nu)(x_{\nu o}x_{\nu 1},x_{\nu 2},\dots,x_{\nu n})$$

and we have proved our claim.

Remark 2.2.3 Let $1 \le r < \infty$ be finite.
Recall (I.2.7.11) the trace homomorphism

$Tr_*: C(M_r(A)) \to C(A)$, given by

$$Tr_n((x_o \otimes a_o),\ldots,(x_n \otimes a_n)) = tr(x_o\cdot\ldots\cdot x_n)(a_o,\ldots,a_n)$$

for $x_i \in M_r(k)$, $a_i \in A$, $0 \le i \le n$.

We also dispose of the domino-indexing formula (I.2.7.12):

$$Tr_n(m^o,m^1,\ldots,m^n) = \sum_{(i_o,i_1,\ldots i_n)} (m^o_{i_o i_1}, m^1_{i_1 i_2},\ldots,m^n_{i_n i_o})$$

Combining I.2.2.6 and I.2.7.14/15, we get a commutative diagram of quasi-isomorphisms

$$\begin{array}{ccc} {}_BC(M_r(A)) & \xrightarrow{Tr_*} & {}_BC(A) \\ \downarrow f & & \downarrow f \\ Tot(C(M_r(A)) & \xrightarrow{Tr_*} & Tot(C(A)) \\ \downarrow \rho & & \downarrow \rho \\ C^\lambda_*(M_r(A)) & \xrightarrow{Tr_*} & C^\lambda_*(A) \end{array}$$

(which is compatible with the standard inclusions $M_r(A) \subset M_{r+1}(A)$).

We have homomorphisms of chain cimplexes

$$Tr_*\circ\lambda_*: (E(gl_r(A))[-1],d[-1]) \to C^\lambda_*(A)$$

which are compatible with the inclusions $gl_r(A) \subset gl_{r+1}(A)$. When passing to the direct limit and taking homology, we obtain finally a homomorphism

$$Tr_*\circ\lambda_*: H_{*+1}(gl(A)) \to HC_*(A)$$

which, on the level of chains, is given by the formula

$$Tr_n\circ\lambda_n((x_o \otimes a_o)\wedge\ldots\wedge(x_n \otimes a_n))$$

$$= (-1)^n \sum_{\sigma\in\gamma_n} \varepsilon(\sigma) tr(x_o\circ x_{\sigma(1)}\circ\ldots\circ x_{\sigma(n)})(a_o,a_{\sigma(1)},\ldots,a_{\sigma(n)}) \bmod (1-t)$$

where $x_o,x_1,\ldots,x_n \in M_\infty(k)$, $a_o,a_1,\ldots,a_n \in A$.

Note that the trace-form $tr: M_\infty(k) \to k$ is well-defined.

<u>Lemma 2.2.4</u> There is a commutative diagram of homomorphisms of chain

complexes

$$\begin{array}{ccc} (E(gl(A)),d) & \xrightarrow{Tr_* \circ \lambda_*} & \\ \downarrow \text{can} & & \\ (E(gl(A))_{\mathfrak{g}},d) & \xrightarrow{Tr_* \circ \lambda_*} & C_*^\lambda(A) \end{array}$$

(we have dropped the dimension-shift notation, but don't forget it ..)

<u>Proof</u>. We have only to show that

$Tr_* \circ \lambda_* \circ \theta(x) = 0$ for all $x \in \mathfrak{g} = \mathfrak{g}_\infty$.

But $\theta(x)((x_o \otimes a_o) \wedge (x_1 \otimes a_1) \wedge \ldots \wedge (x_n \otimes a_n))$

$$= ([x,x_o] \otimes a_o) \wedge (x_1 \otimes a_1) \wedge \ldots \wedge (x_n \otimes a_n)$$

$$+ (x_o \otimes a_o) \wedge ([x,x_1] \otimes a_1) \wedge \ldots \wedge (x_n \otimes a_n) + \ldots$$

$$+ (x_o \otimes a_o) \wedge (x_1 \otimes a_1) \wedge \ldots \wedge ([x,x_n] \otimes a_n)$$

Now, writing out the brackets, it is clear that

$$tr([x,x_o]y_1 \ldots y_n + x_o[x,y_1]y_2 \ldots y_n + \ldots + x_o y_1 \ldots y_{n-1}[x,y_n]) = 0$$

for all $y_1,\ldots,y_n \in M_\infty(k)$.

With $(y_1,\ldots,y_n) = (x_{\sigma 1},\ldots,x_{\sigma n})$, $\sigma \in \gamma_n$, and the explicit formula for $Tr_n \circ \lambda_n$ (cf. 2.2.3), we obtain what we want.

<u>Remark 2.2.5</u>

(1) Observe first that the surjectivity of $Tr_* \circ \lambda_*$ follows from

$$Tr_{n-1} \circ \lambda_{n-1}(E_{12}^{a_1} \wedge E_{23}^{a_2} \wedge \ldots \wedge E_{n1}^{a_n}) = (-1)^{n-1}(a_1,\ldots,a_n) \bmod (1-t)$$

(where $E_{\mu\nu}^{a_\lambda} = E_{\mu\nu} \otimes a_\lambda$ has a_λ as the only nonzero entry on the (μ,ν) position; for $a_\lambda \neq 0$, of course).

The verification is easy:

$$\mathrm{Tr}_{n-1}\circ\lambda_{n-1}(E_{12}^{a_1}\wedge E_{23}^{a_2}\wedge\ldots\wedge E_{n1}^{a_n})$$

$$= (-1)^{n-1}\sum_{\sigma\in\gamma_{n-1}} \varepsilon(\sigma)\,\mathrm{tr}(E_{12}\circ E_{\sigma(2,3)}\circ\ldots\circ E_{\sigma(n,1)})(a_1,a_{\sigma 2},\ldots a_{\sigma n})\,\mathrm{mod}(1-t)$$

$$= (-1)^{n-1}(a_1,\ldots,a_n)\,\mathrm{mod}(1-t)$$

since for $\sigma \neq \mathrm{id}$ $E_{12}\circ E_{\sigma(2,3)}\circ\ldots\circ E_{\sigma(n,1)} = 0$ (think of γ_{n-1} as the permutation group of $\{\underline{2} = (2,3),\ \underline{3} = (3,4),\ldots\ \underline{n} = (n,1)\}$ and of $\{2,3,\ldots n\}$)

(2) Let us next introduce some useful notation: For $a = (a_1,\ldots a_n) \in A^n$ and $\sigma \in \gamma_n$ we define

$$\hat{E}^a_\sigma := E_{1\sigma(1)}^{a_1}\wedge E_{2\sigma(2)}^{a_2}\wedge\ldots\wedge E_{n\sigma(n)}^{a_n}$$

In particular, with $t = (1,2,\ldots n)$, the standard cycle of maximal length in γ_n, we have

$$\hat{E}^a_t = E_{12}^{a_1}\wedge E_{23}^{a_2}\wedge\ldots\wedge E_{n1}^{a_n}.$$

Defining $a\sigma = (a_{\sigma(1)},\ldots a_{\sigma(n)})$, and observing that $\sigma t\sigma^{-1} = (\sigma(1),\ldots \sigma(n))$ (as a cycle), we can easily deduce

$$\hat{E}^{a\sigma^{-1}}_{\sigma t\sigma^{-1}} = \varepsilon(\sigma)E_{\sigma(1)\sigma(2)}^{a_1}\wedge E_{\sigma(2)\sigma(3)}^{a_2}\wedge\ldots\wedge E_{\sigma(n)\sigma(1)}^{a_n}$$

This yields an immediate generalization of (1), by the same arguments; i.e. we get

$$\mathrm{Tr}_{n-1}\circ\lambda_{n-1}(\hat{E}^a_{\sigma t\sigma^{-1}}) = (-1)^{n-1}\varepsilon(\sigma)(a_{\sigma(1)},\ldots a_{\sigma(n)})\,\mathrm{mod}(1-t)$$

(3) Recall now 2.1.12(1), combined with 2.1.8(3): We have an isomorphism

$$(k[\gamma_n]\otimes A^n)\underset{\gamma_n}{\otimes}(\mathrm{sgn}) \xrightarrow{\sim} E_n(\mathcal{g}\otimes A)_{\mathcal{g}}$$

$$\sigma \otimes (a_1,\dots,a_n) \otimes 1 \mapsto \sum_{i\in[\underline{n},\underline{n}]} \hat{E}^{a}_{i,i\sigma} \bmod \mathfrak{g}E_n(\mathfrak{g} \otimes A)$$

In order to streamline the arguments in the sequel, we have to get rid of the long sum-expression, which is redundant in the following sense:

<u>Claim</u>. $(k[\gamma_n] \otimes A^n) \underset{\gamma_n}{\otimes} (\mathrm{sgn}) \xrightarrow{\sim} E_n(\mathfrak{g} \otimes A)_{\mathfrak{g}}$

via $\quad \sigma \otimes (a_1,\dots,a_n) \otimes 1 \mapsto \hat{E}^{a}_{\sigma} \bmod \mathfrak{g}E_n(\mathfrak{g} \otimes A)$

(Note that this is <u>not</u> a mere reduction of the sum expression above modulo $\mathfrak{g}E(\mathfrak{g} \otimes A)$).

<u>Proof</u>. It is sufficient to show the following:
Put (exceptionally) $\mathfrak{g} = \mathfrak{gl}_n(k)$. Then $k[\gamma_n]$ is k-isomorphic to $(\mathfrak{g}^n)_{\mathfrak{g}}$ via $\sigma \mapsto E_{id,\sigma} \bmod \mathfrak{g}\cdot\mathfrak{g}^n$. (where $E_{id,\sigma} = E_{1,\sigma(1)} \otimes \dots \otimes E_{n,\sigma(n)}$).
We have to make sure that the set $\{E_{id,\sigma}: \sigma \in \gamma_n\}$ is k-linearly independent modulo $\mathfrak{g}\cdot\mathfrak{g}^n$. This is easily seen by a duality argument. Consider $W = \mathfrak{g}^n$, and identify $W^* = \mathrm{Hom}_k(\mathfrak{g}^n,k)$ with $\mathfrak{g}^n$ via the scalar product

$$\langle E_{ij},E_{k\ell}\rangle = \begin{cases} 1 & \text{if } \quad i = \ell \text{ and } j = k \\ 0 & \text{else} \end{cases}$$

(for $i,j,k,\ell \in [\underline{n},\underline{n}]$).
Recall now 2.1.8: The duality above restricts to a duality of $(\mathfrak{g}^{n*})^{\mathfrak{g}} = (\mathfrak{g}^n)^{\mathfrak{g}}$ with $(\mathfrak{g}^n)_{\mathfrak{g}}$.
Consider $\underline{\tau} = \sum_{i\in[\underline{n},\underline{n}]} E_{i\tau,i} \in (\mathfrak{g}^n)^{\mathfrak{g}}$ and $\underline{\sigma}' = E_{id,\sigma} \in \mathfrak{g}^n$ (or $(\mathfrak{g}^n)_{\mathfrak{g}}$).
We get:

$$\langle\underline{\tau},\underline{\sigma}'\rangle = \sum_{i\in[\underline{n},\underline{n}]} \langle E_{i\tau,i},E_{id,\sigma}\rangle = \begin{cases} 1 & \text{if } \quad \tau = \sigma \\ 0 & \text{else} \end{cases}$$

which shows finally our claim.

(4) Let us calculate $\mathrm{Tr}_{n-1}\circ\lambda_{n-1}(t \otimes (a_1,\dots,a_n) \otimes 1)$ via the reduced isomorphism (3):

$$\begin{aligned} \mathrm{Tr}_{n-1}\circ\lambda_{n-1}(t \otimes (a_1,\dots,a_n) \otimes 1) &= \mathrm{Tr}_{n-1}\circ\lambda_{n-1}(\hat{E}^{a}_{t}) \\ &= (-1)^{n-1}(a_1,\dots,a_n) \bmod (1-t) \end{aligned}$$

Since $\sigma t \sigma^{-1} \otimes (a_1,\ldots,a_n) \otimes 1 = t \otimes (a_{\sigma(1)},\ldots,a_{\sigma(n)}) \otimes \varepsilon(\sigma)$ we get immediately (cf. (2)):

$$Tr_{n-1} \circ \lambda_{n-1} (\sigma t \sigma^{-1} \otimes (a_1,\ldots,a_n) \otimes 1) = (-1)^{n-1} \varepsilon(\sigma)(a_{\sigma(1)},\ldots,a_{\sigma(n)}) \mathrm{mod}(1-t)$$

Remark-Definition 2.2.6

(1) An explicit calculation within $(E(gl(A)),d)$ yields

$$d(E_{12}^{a_1} \wedge E_{23}^{a_2} \wedge\ldots\wedge E_{n1}^{a_n})$$

$$= \sum_{i=1}^{n-1} (-1)^i E_{12}^{a_1} \wedge\ldots\wedge E_{i,i+2}^{a_i a_{i+1}} \wedge\ldots\wedge E_{n1}^{a_n}$$

$$+ (-1)^n E_{n,2}^{a_n a_1} \wedge\ldots\wedge E_{n-1,n}^{a_{n-1}}$$

and, more generally:

$$d(E_{\sigma(1)\sigma(2)}^{a_1} \wedge\ldots\wedge E_{\sigma(n)\sigma(1)}^{a_n})$$

$$= \sum_{i=1}^{n-1} (-1)^i E_{\sigma(1)\sigma(2)}^{a_1} \wedge\ldots\wedge E_{\sigma(i)\sigma(i+2)}^{a_i a_{i+1}} \wedge\ldots\wedge E_{\sigma(n)\sigma(1)}^{a_n}$$

$$+ (-1)^n E_{\sigma(n)\sigma(2)}^{a_n a_1} \wedge E_{\sigma(2)\sigma(3)}^{a_2} \wedge\ldots\wedge E_{\sigma(n-1)\sigma(n)}^{a_{n-1}}$$

(with the convention: $n+1 = 1$, whenever it appears)

(2) We shall define a subcomplex $(\mathbb{P}E(gl(A)),d)$ of $(E(gl(A)),d)$:

$\mathbb{P}_n E(gl(A)) \subset E_n(gl(A))$ is the k-subspace of $E_n(g \otimes A)$ which is spanned by all injectively domino-indexed elements

$E_{i_1 i_2}^{a_1} \wedge E_{i_2 i_3}^{a_2} \wedge\ldots\wedge E_{i_n i_1}^{a_n}$, with $i = (i_1,\ldots,i_n) \in \varinjlim_r [\underline{r},\underline{n}]$ one-to-one, $(a_1,\ldots,a_n) \in A^n$.

By virtue of (1) we obtain actually a subcomplex of $(E(gl(A)),d)$.

(3) Define the subcomplex (of primitive elements)

$$\mathbb{P}(E(gl(A))_g) = \mathrm{Im}(\mathbb{P}E(gl(A)) \to E(gl(A))_g)$$

<u>Claim</u>. $\mathbb{P}_n E(gl(A))_g) = \{E_{12}^{a_1} \wedge \ldots \wedge E_{n1}^{a_n} \bmod gE_n(g \otimes A) : a_1, \ldots, a_n \in A\}$

<u>Proof</u>. First, $\mathbb{P}_n(E(gl(A))_g)$ is the k-linear span of all $E_{\sigma(1)\sigma(2)}^{a_1} \wedge E_{\sigma(2)\sigma(3)}^{a_2} \wedge \ldots \wedge E_{\sigma(n)\sigma(1)}^{a_n}$, with $\sigma \in \gamma_n$, $a_1, \ldots, a_n \in A$ (we don't write equivalence classes; note that the important fact is the possibility to restrict to matrices out of $gl_n(A)$).
This follows from 2.1.8 and 2.1.12:

$$E_n(g \otimes A)_g \simeq ((g_n^n)_{g_n} \otimes A^n) \underset{\gamma_n}{\otimes} (\mathrm{sgn})$$

Now, the adjoint action of $\Gamma = GL(n,k)$ (by parallel conjugation) on $E_n(g \otimes A)_g$ is <u>trivial</u>. But

$$E_{\sigma(1)\sigma(2)}^{a_1} \wedge \ldots \wedge E_{\sigma(n)\sigma(1)}^{a_n} = g(\sigma).E_{12}^{a_1} \wedge \ldots \wedge E_{n1}^{a_n}$$

where $g(\sigma) = \sum_{j=1}^{n} E_{\sigma(j),j} \in \Gamma = GL(n,k)$ is the permutation matrix corresponding to $\sigma \in \gamma_n$.

<u>Lemma 2.2.7</u> Let $U_n \subset \gamma_n$ be the conjugacy class of $t = (1,2,\ldots,n)$. Then the reduced isomorphism 2.2.5(3) induces an isomorphism

$$(k[U_n] \otimes A^n) \underset{\gamma_n}{\otimes} (\mathrm{sgn}) \to \mathbb{P}_n(E(g \otimes A)_g), \quad n \geq 1$$

<u>Proof</u>. Recall the explicit form of the reduced isomorphism

$$(k[\gamma_n] \otimes A^n) \underset{\gamma_n}{\otimes} (\mathrm{sgn}) \to E_n(g \otimes A)_g$$

$$\sigma \otimes (a_1, \ldots, a_n) \otimes 1 \mapsto \hat{E}_\sigma^a \bmod gE_n(g \otimes A)$$

Hence $\sigma t \sigma^{-1} \otimes (a_1, \ldots, a_n) \otimes 1 = t \otimes (a_{\sigma(1)}, \ldots, a_{\sigma(n)}) \otimes \varepsilon(\sigma)$ identifies with the class of $\hat{E}^a_{\sigma t \sigma^{-1}} = \varepsilon(\sigma) E_{\sigma(1)\sigma(2)}^{a_{\sigma(1)}} \wedge \ldots \wedge E_{\sigma(n)\sigma(1)}^{a_{\sigma(n)}}$ (which equals the class of $\hat{E}_t^{a\sigma} = \varepsilon(\sigma) E_{12}^{a_{\sigma(1)}} \wedge \ldots \wedge E_{n1}^{a_{\sigma(n)}}$; cf. the argument at the end of 2.2.6(3)).
The assertion of the lemma is now immediate.

We get the following commutative diagram

$$\begin{array}{ccc} \mathbb{P}_n(E(g \otimes A)_g) & \xrightarrow{\sim} & (k[U_n] \otimes A^n) \underset{\gamma_n}{\otimes} (\mathrm{sgn}) \\ & {\scriptstyle Tr_{n-1}\circ\lambda_{n-1}} \searrow \quad \swarrow {\scriptstyle Tr_{n-1}\circ\lambda_{n-1}} & \\ & C^{\lambda}_{n-1}(A) = (A^n)/(1-t) & \end{array}$$

where, as already shown,

$$Tr_{n-1}\circ\lambda_{n-1}(\sigma t\sigma^{-1} \otimes (a_1,\ldots,a_n) \otimes 1)$$

$$= (-1)^{n-1}\varepsilon(\sigma)(a_{\sigma(1)},\ldots,a_{\sigma(n)})\mathrm{mod}(1-t)$$

Proposition 2.2.8

$$Tr_*\circ\lambda_*: \ \mathbb{P}(E(g \otimes A)_g)[-1] \to C^{\lambda}_*(A)$$

is an isomorphism of chain complexes.

Proof. It is sufficient to show that

$$Tr_{n-1}\circ\lambda_{n-1}: (k[U_n]\otimes A^n) \underset{\gamma_n}{\otimes} (\mathrm{sgn}) \to C^{\lambda}_{n-1}(A) = A^n/(1-t)$$

is an isomorphism for all $n \geq 1$.
We shall exhibit a composition of several isomorphisms which finally maps $t \otimes (a_1,\ldots,a_n) \otimes 1$ to $(a_1,\ldots,a_n)\mathrm{mod}(1-t)$, thus establishing our claim.
First, identify $\mathbb{Z}/n\mathbb{Z}$ with $G_n = \langle t\rangle \subset \gamma_n$, where $t = (1,2,..,n)$.
Then $U_n = \{\sigma t\sigma^{-1} = (\sigma(1),\ldots,\sigma(n)),\ \sigma \in \gamma_n\}$ identifies with γ_n/G_n $= \{\sigma G_n: \sigma \in \gamma_n\}$ via $\sigma t\sigma^{-1} \to \mathrm{cl}(\sigma) = \sigma G_n$ as γ_n-sets: The left action of γ_n on U_n by conjugation corresponds to the left action of γ_n on γ_n/G_n by left multiplication.
Consider now k as a G_n-module, with trivial G_n-action. Then $k[\gamma_n/G_n] \simeq k[\gamma_n] \underset{G_n}{\otimes} k$, which yields an isomorphism

$$(k[U_n] \otimes A^n) \underset{\gamma_n}{\otimes} (\mathrm{sgn}) \simeq (k[\gamma_n] \underset{G_n}{\otimes} A^n) \underset{\gamma_n}{\otimes} (\mathrm{sgn})$$

mapping $\sigma t\sigma^{-1} \otimes (a_1,\ldots,a_n) \otimes 1$ to $\sigma \otimes (a_1,\ldots,a_n) \otimes 1$.

Let us continue:

$$(k[\gamma_n] \otimes_{G_n} A^n) \otimes_{\gamma_n} (\text{sgn}) \simeq A^n \otimes_{G_n} (\text{sgn}) \quad \text{via}$$

$$\sigma \otimes (a_1,\dots,a_n) \otimes 1 = \text{id} \otimes (a_{\sigma^{-1}(1)},\dots,a_{\sigma^{-1}(n)}) \otimes \varepsilon(\sigma) \to$$

$$(a_{\sigma^{-1}(1)},\dots,a_{\sigma^{-1}(n)}) \otimes \varepsilon(\sigma)$$

and $A^n \otimes_{G_n} (\text{sgn}) \simeq A^n/(1-t)$ by means of the identification $(a_1,\dots,a_n) \otimes 1 = (a_1,\dots,a_n) \bmod (1-t)$. It is easily verified that the image of $t \otimes (a_1,\dots,a_n) \otimes 1$ by the isomorphism $(k[U_n] \otimes A^n) \otimes_{\gamma_n} (\text{sgn}) \simeq A^n/(1-t)$ is precisely $(a_1,\dots,a_n) \bmod (1-t)$.

This finishes the proof of our proposition.

<u>Definition 2.2.9</u> The inclusion $\mathbb{P}E(gl(A)) \overset{i}{\subset} E(gl(A))$ defines in homology the k-vector space

$$\text{Prim } H_*(gl(A)) = \text{Im } H_*(i) \subset H_*(gl(A))$$

of <u>primitive elements</u> of $H_*(gl(A))$.

<u>Theorem 2.2.10</u>

$Tr_* \circ \lambda_*$: Prim $H_*(gl(A))[-1] \to HC_*(A)$ is an isomorphism (of graded k-vector spaces).

<u>Proof</u>. This follows immediately from I.2.2.6, combined with 2.1.12 and 2.2.8. Note that we dispose of a commutative diagram of complex homomorphisms

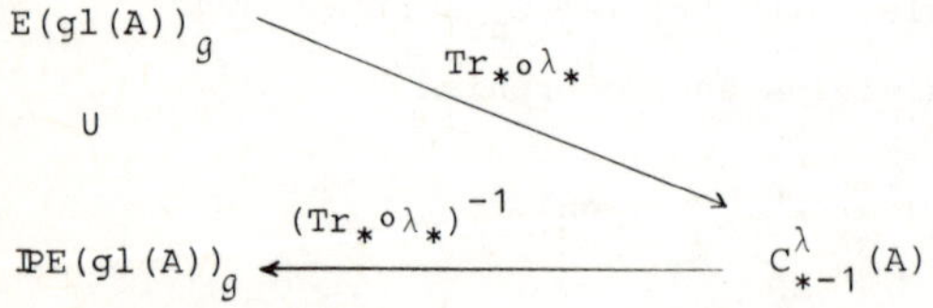

$$\begin{array}{ccc} E(gl(A))_g & \xrightarrow{Tr_* \circ \lambda_*} & \\ \cup & & C^\lambda_{*-1}(A) \\ \mathbb{P}E(gl(A))_g & \xleftarrow{(Tr_* \circ \lambda_*)^{-1}} & \end{array}$$

which yields a projection of $E(gl(A))_g$ onto $\mathbb{P}E(gl(A))_g$.

In order to bring this section to a satisfactory end, let us sketch the significance of primitivity for the homology of $gl(A)$.
For later reference, we need the following

Lemma 2.2.11 Consider $\alpha \in GL(r,k)$, $r \geq 1$, as an element of $GL(k) = \varinjlim GL(r,k)$.
Conjugation with α defines a k-automorphism of $gl(A)$, which translates to the usual adjoint action of α on $E(g \otimes A)$, by parallel conjugation on the g-components. The induced action of α on $E(g \otimes A)_g$ is trivial, hence the induced action of α on $H_*(gl(A))$ is trivial, too.

Proof. The action of $g_n = gl_n(k)$ on $E_n(g \otimes A)_g$ is trivial, and it is the differential of the adjoint action of $\Gamma_n = GL(n,k)$; cf. [Hu, p.88].

Remark 2.2.12 The structure of $H_*(gl(A))$ as a graded Hopf-algebra.
(1) Recall the definition of a graded Hopf-algebra H over a commutative ring k (cf. [ML, pp.197-200]): $H = (H_n)_{n\geq 0}$ is a graded k-module, which, with this grading, is both a graded k-algebra for a product map $\mu: H \otimes H \to H$ and a unit $\eta: k \to H$, and a graded k-coalgebra for a diagonal $\Delta: H \to H \otimes H$ and a counit $\varepsilon: H \to k$ such that

(i) $\eta: k \to H$ is a homomorphism of graded coalgebras

(ii) $\varepsilon: H \to k$ is a homomorphism of graded algebras

(iii) $\Delta: H \to H \otimes H$ is a homomorphism of graded algebras

A typical example, closest to the set-up of this section, is the exterior algebra $E(V) = \bigoplus_{n\geq 0} \Lambda^n V$ of a k-module V.
The structure as a graded k-algebra is clear. $\varepsilon: E(V) \to k$ is the projection on $E_0(V) = k$, and $\Delta: E(V) \to E(V) \otimes E(V)$ (tensor product of graded algebras!) is induced by the diagonal map $\Delta: V \to V \times V$, which prolongs to a homomorphism of graded k-algebras

$$\Delta: E(V) \to E(V \times V) = E(V) \otimes E(V).$$

Explicitely, we have the following formula:

$\Delta: E_n(V) \to \sum_{p+q=n} E_p(V) \otimes E_q(V)$ is given by

$\Delta(x_1 \wedge \ldots \wedge x_n) = \sum_{p+q=n} \Delta_{p,q}(x_1 \wedge \ldots \wedge x_n)$, where

$$\Delta_{p,q}(x_1 \wedge \ldots \wedge x_n) = \sum_{(\sigma)} \varepsilon(\sigma)(x_{\sigma(1)} \wedge \ldots \wedge x_{\sigma(p)}) \otimes (x_{\sigma(p+1)} \wedge \ldots \wedge x_{\sigma(n)})$$

and the sum is taken over all $\sigma \in \gamma_n$ which are increasing on either interval $[1,p]$ and $[p+1,n]$.

(2) Intermediate observation:

Let (g_1, g_2) be a couple of Lie algebras over a <u>field</u> k. We want to indentify the homology $H_*(g_1 \times g_2)$ of the direct product $g_1 \times g_2$ with $H_*(g_1) \otimes H_*(g_2)$ in the following sense:

$$H_n(g_1 \times g_2) \simeq \sum_{p+q=n} H_p(g_1) \otimes H_q(g_2), \quad n \geq 0.$$

This follows from the Künneth-formula (cf. [ML,p.166]), provided we have made sure that $E(g_1 \times g_2)$ identifies with the tensor-product $E(g_1) \otimes E(g_2)$ as a chain complex, too. First, we have the isomorphism of graded k-algebras $E(g_1 \times g_2) \simeq E(g_1) \otimes E(g_2)$ which reads in degree n like this:

$$E_n(g_1 \times g_2) \overset{\sim}{\to} \sum_{p+q=n} E_p(g_1) \otimes E_q(g_2)$$

$$(x_1,y_1) \wedge \ldots \wedge (x_n,y_n) \mapsto \sum_{p+q=n} \sum_{(\sigma)} \varepsilon(\sigma)(x_{\sigma(1)} \wedge \ldots \wedge x_{\sigma(p)}) \otimes (y_{\sigma(p+1)} \wedge \ldots \wedge y_{\sigma(n)})$$

(where the (σ)-summation is as at the end of (1)).

The transport of the differential from the left to the right gives precisely the usual differential of the tensor product of (chain) complexes. Once more (read in the other direction):

$$[x_1 \wedge \ldots \wedge x_p] \otimes [y_1 \wedge \ldots \wedge y_q] \in H_p(g_1) \otimes H_q(g_2)$$

identifies with the homology class of

$$(x_1,0) \wedge (x_2,0) \wedge \ldots \wedge (x_p,0) \wedge (0,y_1) \wedge (0,y_2) \wedge \ldots \wedge (0,y_q)$$

in $H_{p+q}(g_1 \times g_2)$.

(3) Let now A be a unital associative algebra over a field k of characteristic 0.

Consider $H = H_*(gl(A))$, as a graded k-vector space. We aim at making explicit a (graded) k-algebra structure as well as a (graded) k-coalgebra structure on H such that H becomes a (graded) Hopf-algebra.

Let us look first at the coalgebra structure.

The counit $\varepsilon : H \to k$ is merely the projection on the first factor $H_0 = H_0(gl(A)) = k$. As to the comultiplication $\Delta : H \to H \otimes H$, it is induced by the diagonal map $\Delta : gl(A) \to gl(A) \times gl(A)$ which is a homomorphism of Lie algebras, hence gives rise to $\Delta: H_*(gl(A)) \to H_*(gl(A)\times gl(A))$. Now, according to (2), we can identify $H_*(gl(A) \times gl(A))$ with $H_*(gl(A)) \otimes H_*(gl(A))$, such that Δ is given explicitely by the following formula (brackets mean homology classes):

$$\Delta([x_1\wedge\ldots\wedge x_n]) = \sum_{p+q=n} \sum_{(\sigma)} \varepsilon(\sigma)[x_{\sigma(1)}\wedge\ldots\wedge x_{\sigma(p)}]\otimes[x_{\sigma(p+1)}\wedge\ldots\wedge x_{\sigma(n)}]$$

(with our familiar extra-condition on the $\sigma \in \gamma_n$).

So far the particular structure of $gl(A)$ is not needed. As to the graded algebra structure of H, the unit $\eta: k \to H$ is given by the inclusion of $k = H_0$ in H. In order to obtain the multiplication, look first at the family of Lie algebra homomorphisms

$$\mu_r: gl_r(A) \times gl_r(A) \to gl_{2r}(A) \qquad r \geq 1$$

$$(x,y) \to \begin{pmatrix} x & 0 \\ 0 & y \end{pmatrix}$$

Note that the μ_r, $r \geq 1$, are <u>not</u> compatible with the standard inclusions $gl_r(A) \subset gl_{r+1}(A)$. But they <u>are</u> compatible up to conjugation. More precisely,

$$\text{let } \alpha_r = \begin{pmatrix} Id_r & 0 & 0 & 0 \\ 0 & 0 & 1 & 0 \\ 0 & Id_r & 0 & 0 \\ 0 & 0 & 0 & 1 \end{pmatrix} \in GL(2r+2,k).$$

Then for $(x,y) \in gl_r(A) \times gl_r(A) \subset gl_{r+1}(A) \times gl_{r+1}(A)$ we have:

$$\mu_r(x,y) = \alpha_r \mu_{r+1}(x,y)\alpha_r^{-1}$$

(relative to $gl_{2r}(A) \subset gl_{2r+2}(A)$).

Thus, by 2.2.11, we can pass to the direct limit <u>in homology</u> and get

$$\mu: H_*(gl(A)) \otimes H_*(gl(A)) = H_*(gl(A) \times gl(A)) \to H_*(gl(A)).$$

More explicitely,

$$\mu: \sum_{p+q=n} H_p(gl(A)) \otimes H_q(gl(A)) \to H_n(gl(A)), \quad n \geq 0,$$

is given by the following formula:

For $[x_1\wedge\ldots\wedge x_p] \otimes [y_1\wedge\ldots\wedge y_q] \in H_p(gl(A)) \otimes H_q(gl(A))$

$\mu([x_1\wedge\ldots\wedge x_p] \otimes [y_1\wedge\ldots\wedge y_q]) \in H_{p+q}(gl(A))$ is the homology class of $\begin{pmatrix} x_1 & 0 \\ 0 & 0 \end{pmatrix} \wedge\ldots\wedge \begin{pmatrix} x_p & 0 \\ 0 & 0 \end{pmatrix} \wedge \begin{pmatrix} 0 & 0 \\ 0 & y_1 \end{pmatrix} \wedge\ldots\wedge \begin{pmatrix} 0 & 0 \\ 0 & y_q \end{pmatrix}$ where, for $x_1,\ldots,x_p$, $y_1,\ldots,y_q \in gl_r(A)$, say, the exterior product above is now in $E_{p+q}(gl_{2r}(A))$. $H = H_*(gl(A))$ becomes thus a graded unital associative k-algebra.

The verification of the required properties for the graded algebra structure as well as for the graded coalgebra structure of H is rather straightforward. As to the compatibility properties, which give finally the graded Hopf-algebra structure, the only delicate point is the verification of the fact that the comultiplication $\Delta:H \to H \otimes H$ is a homomorphism of graded k-algebras.

Let us look at the situation.

Take $u = [x_1\wedge\ldots\wedge x_p] \in H_p(gl(A))$, $v = [y_1\wedge\ldots\wedge y_q] \in H_q(gl(A))$.

We have to show that

$$\Delta(u)\Delta(v) = \Delta(\mu(u \otimes v)) \quad \text{in} \quad (H \otimes H)_{p+q}.$$

(the multiplication on the left side is in the graded tensor product $H \otimes H$)

$$\Delta(u) = [(x_1 \otimes 1 + 1 \otimes x_1)\ldots(x_p \otimes 1 + 1 \otimes x_p)]$$

$$\Delta(v) = [(y_1 \otimes 1 + 1 \otimes y_1)\ldots(y_q \otimes 1 + 1 \otimes y_q)]$$

(the product inside the brackets is the product of the graded tensor product $E(gl(A)) \otimes E(gl(A))$

$\mu(u \otimes v)$ is represented by

$$\begin{pmatrix} x_1 & 0 \\ 0 & 0 \end{pmatrix} \wedge\ldots\wedge \begin{pmatrix} x_p & 0 \\ 0 & 0 \end{pmatrix} \wedge \begin{pmatrix} 0 & 0 \\ 0 & y_1 \end{pmatrix} \wedge\ldots\wedge \begin{pmatrix} 0 & 0 \\ 0 & y_q \end{pmatrix}; \quad \text{consequently}$$

$$\Delta(\mu(u \otimes v)) = \left[\left(\begin{pmatrix} x_1 & 0 \\ 0 & 0 \end{pmatrix} \otimes 1 + 1 \otimes \begin{pmatrix} x_1 & 0 \\ 0 & 0 \end{pmatrix}\right)\ldots\left(\begin{pmatrix} 0 & 0 \\ 0 & y_q \end{pmatrix} \otimes 1 + 1 \otimes \begin{pmatrix} 0 & 0 \\ 0 & y_q \end{pmatrix}\right)\right]$$

Now look at $\Delta(u)\Delta(v)$ in $(H \otimes H)_{p+q}$.

The multiplication in $H \otimes H$ is given by the following formula:

$$([a_1\wedge\cdots\wedge a_r] \otimes [b_1\wedge\cdots\wedge b_s])\cdot([a'_1\wedge\cdots\wedge a'_{r'}] \otimes [b'_1\wedge\cdots\wedge b'_{s'}]) =$$

$$= (-1)^{s\cdot r'}\left[\begin{pmatrix} a_1 & 0 \\ 0 & 0\end{pmatrix}\wedge\cdots\wedge\begin{pmatrix} 0 & 0 \\ 0 & a'_{r'}\end{pmatrix}\right] \otimes \left[\begin{pmatrix} b_1 & 0 \\ 0 & 0\end{pmatrix}\wedge\cdots\wedge\begin{pmatrix} 0 & 0 \\ 0 & b'_{s'}\end{pmatrix}\right]$$

This implies:

(i) $\Delta\mu([x] \otimes [y]) = \Delta([x])\Delta([y])$

$$= \left[\begin{pmatrix} x & 0 \\ 0 & 0\end{pmatrix} \wedge \begin{pmatrix} 0 & 0 \\ 0 & y\end{pmatrix} \otimes 1 + 1 \otimes \begin{pmatrix} x & 0 \\ 0 & 0\end{pmatrix} \wedge \begin{pmatrix} 0 & 0 \\ 0 & y\end{pmatrix}\right]$$

for $x, y \in gl(A)$.

(ii) With $\bar{u} = u_1 \wedge\cdots\wedge u_r$, $\bar{v} = v_1 \wedge\cdots\wedge v_s$

$$(\bar{u},0) = \begin{pmatrix} u_1 & 0 \\ 0 & 0\end{pmatrix} \wedge\cdots\wedge \begin{pmatrix} u_r & 0 \\ 0 & 0\end{pmatrix}$$

$$(0,\bar{v}) = \begin{pmatrix} 0 & 0 \\ 0 & v_1\end{pmatrix} \wedge\cdots\wedge \begin{pmatrix} 0 & 0 \\ 0 & v_s\end{pmatrix}$$

we have: $\Delta\mu([\bar{u}\wedge x]\otimes[\bar{v}]) = \Delta([\bar{u}\wedge x])\Delta([\bar{v}]) = [\Delta((\bar{u},0))\Delta\begin{pmatrix} x & 0 \\ 0 & 0\end{pmatrix}\Delta((0,\bar{v}))]$

(where the last product inside the brackets is in $E(gl(A)) \otimes E(gl(A))$).

Analogously for the other side.
Our result follows by induction on $n = p+q$.

(4) $H = H_*(gl(A))$ is a connected commutative and cocommutative graded Hopf-algebra.

(a) Connectedness means merely that $H_0 = H_0(gl(A)) = k$.

(b) Commutativity (in the graded sense) means that

$$h_p h_q = (-1)^{pq} h_q h_p \quad \text{for all } h_p \in H_p,\ h_q \in H_q,\ p,q \geq 0.$$

This follows immediately from the definition of the multiplication (and from Lemma 2.2.11), since, for $x,y \in gl_r(A)$, $\mu_r(x,y)$ and $\mu_r(y,x)$ are conjugate by $\xi_r = \begin{pmatrix} 0 & Id_r \\ -Id_r & 0\end{pmatrix} \in GL(2r,k)$.

(c) Cocommutativity means the following:

Let $T : H \otimes H \to H \otimes H$ be the twisting morphism given by

$$T_n(h_p \otimes h_q) = (-1)^{pq} h_q \otimes h_p \quad \text{for } h_p \in H_p,\ h_q \in H_q,\ p+q = n.$$

Then the following diagram is commutative:

$$\begin{array}{ccc} & \xrightarrow{\Delta} & H \otimes H \\ H & & \downarrow T \\ & \xrightarrow{\Delta} & H \otimes H \end{array}$$

This property is already valid on the chain-level:

$T: E(gl(A)) \otimes E(gl(A)) \rightarrow E(gl(A)) \otimes E(gl(A))$

is an automorphism of the graded k-algebra $E(gl(A)) \otimes E(gl(A))$; furthermore,

$\Delta: E(gl(A)) \rightarrow E(gl(A)) \otimes E(gl(A))$

is a homomorphism of graded k-algebras, and $\Delta(x) = x \otimes 1 + 1 \otimes x = T\Delta(x)$ for all $x \in gl(A)$.

<u>Remark 2.2.13</u> The primitive part of $H_*(gl(A))$.

(1) Preliminaries.
Let k be a field of characteristic 0, and let H be any connected graded Hopf-algebra over k.
Consider $I(H) = \operatorname{Ker} \varepsilon = \bigoplus_{n\geq 1} H_n$, and look at the two exact sequences

$$I(H) \otimes I(H) \xrightarrow{\mu} I(H) \rightarrow Q(H) \rightarrow 0$$

$$0 \rightarrow P(H) \rightarrow I(H) \xrightarrow{\Delta} I(H) \otimes I(H)$$

which define $P(H)$ and $Q(H)$. The elements of $Q(H)$ are called the <u>indecomposable</u> elements of H, whereas the elements of $P(H)$ are called the <u>primitive</u> elements of H.
More explicitely, we have

$$Q(H) = H_1 \oplus H_2/H_1 \circ H_1 \oplus H_3/H_1 \circ H_2 + H_2 \circ H_1 \oplus \ldots$$

$$P(H) = \{x \in H: \Delta(x) = x \otimes 1 + 1 \otimes x\}.$$

There is a natural homomorphism $P(H) \rightarrow Q(H)$, which is an isomorphism whenever H is commutative and cocommutative (cf. [MM,4.18,p.234]).

(2) The graded Lie algebra structure of $P(H)$.
Let H be a graded Hopf-algebra.
Define the graded commutator $[,]: H \otimes H \rightarrow H$ by $[h_p, h_q] = h_p h_q - (-1)^{pq} h_q h_p$ for $h_p \in H_p$, $h_q \in H_q$, $p,q \geq 0$.

$(H,[\,,\,])$ becomes a graded Lie algebra in the following sense:

(i) $[h_p,h_q] = -(-1)^{pq}[h_q,h_p]$

(ii) $(-1)^{pr}[h_p,[h_q,h_r]] + (-1)^{qp}[h_q,[h_r,h_p]] + (-1)^{rq}[h_r,[h_p,h_q]] = 0$

(for homogeneous elements of the indicated degrees)

Claim. $(P(H),[\,,\,])$ is a graded Lie subalgebra of $(H,[\,,\,])$.
We have show that $[x_p,y_q] \in P(H)_{p+q}$ for $x_p \in P(H)_p$, $y_q \in P(H)_q$.

Now, $\Delta(x_py_q) = \Delta(x_p)\Delta(y_q) = (x_p \otimes 1 + 1 \otimes x_p)(y_q \otimes 1 + 1 \otimes y_q) =$

$$= x_py_q \otimes 1 + 1 \otimes x_py_q + x_p \otimes y_q + (-1)^{pq}y_q \otimes x_p.$$

Thus $\Delta([x_p,y_q]) = [x_p,y_q] \otimes 1 + 1 \otimes [x_p,y_q]$.

Consequence. Let A be a unital associative algebra over a field k of characteristic 0. Consider the graded (connected, commutative and cocommutative) Hopf-algebra $H = H_*(gl(A))$. Its primitive part $P(H)$ is an abelian graded Lie algebra (with $[\,,\,] \equiv 0$).

Remark 2.2.14 Reconstruction of a graded (connected, commutative and cocommutative) Hopf-algebra H over a field k of characteristic 0 from its primitive part $P(H)$.

(a) Let $V = \bigoplus_{K\geq 1} V_K$ be any (positively) graded k-vector space.
Consider the tensor algebra $T(V)$ of V over k, with the following grading:

$$T_o(V) = k \qquad T_n(V) = V_n \oplus (V \otimes V)_n \oplus (V \otimes V \otimes V)_n \oplus \ldots \qquad n \geq 1$$

where $(V^{\otimes m})_n = \sum_{K_1+K_2+\ldots+K_m=n} V_{K_1} \otimes V_{K_2} \otimes \ldots \otimes V_{K_m}$.

Put $\Lambda(V) = T(V)/I$, where I is the two-sided ideal in $T(V)$ which is generated by all graded commutators $x_p \otimes y_q - (-1)^{pq}y_q \otimes x_p$, $x_p \in V_p$, $y_q \in V_q$, $p,q \geq 1$. Since I is homogeneous (for the total grading of $T(V)$), $\Lambda(V)$ inherits a quotient grading such that the canonical map $T(V) \to \Lambda(V)$ is a homomorphism of graded k-algebras. Furthermore, $k \oplus V \to \Lambda(V)$ is a monomorphism of graded k-vector spaces (I is generated by quadratic elements, for the usual grading of $T(V)$). $\Lambda(V)$ is the free graded commutative k-algebra on V in the following sense:

For every commutative graded k-algebra B and every homomorphism of graded k-vector spaces $f: V \to B$ there is a unique homomorphism of graded k-algebras $\tilde{f}: \Lambda(V) \to B$ such that

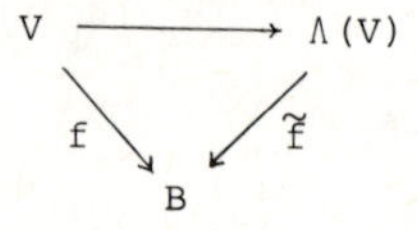

is commutative.

The structure of $\Lambda(V)$ is easy to describe.
Consider first two particular cases:

(i) $V_{2\nu} = 0$ for all $\nu \geq 1$.

Then $\Lambda(V) = E(V)$, the exterior algebra of V over k.

(ii) $V_{2\nu+1} = 0$ for all $\nu \geq 0$.

Then $\Lambda(V) = S(V)$, the symmetric algebra of V over k. (with, in both cases, a total grading coming from the interior grading of V, in analogy with the grading of $T(V)$ above).
Since $\Lambda(V \oplus W) = \Lambda(V) \otimes \Lambda(W)$ (graded tensor product), we obtain immediately the general case:

Decompose $V = V_+ \oplus V_-$ with $V_+ = \bigoplus_{K \text{ even}} V_K$, $V_- = \bigoplus_{K \text{ odd}} V_K$.

Then $\Lambda(V) = S(V_+) \otimes E(V_-)$

Furthermore, $S(V_+) = \bigotimes_{K \text{ even}} S(V_K)$, $E(V_-) = \bigotimes_{K \text{ odd}} E(V_K)$

(all tensor products are graded tensor products of graded k-algebras (direct limits!)).
Note that $\Lambda^n(V_K) = 0$ for $n \not\equiv 0 \mod K$.

(b) Let now L be a (positively) graded abelian Lie algebra over a field k of characteristic 0. Since L is merely a graded k-vector space V without extra-structure, its universal enveloping algebra $U(L)$ (in the graded sense) equals $\Lambda(V)$ (cf. [MM] for a detailed discussion of $U(L)$ in the general case). We shall nevertheless write $U(L)$ in our particular setting, too, in order to accentuate the context of ideas.
$U(L)$ is a connected commutative and cocommutative (graded) Hopf-algebra.
This is clear, once you have observed the following:
The diagonal $\Delta: L \to L \times L$, which is a homomorphism of graded (abelian) Lie algebras, prolongs to the comultiplication $\Delta: U(L) \to U(L \times L) = U(L) \otimes U(L)$, which is a homomorphism of graded k-algebras. The twisting

morphism $T: U(L) \otimes U(L) \to U(L) \otimes U(L)$ is an automorphism of the graded algebra $U(L) \otimes U(L)$. Thus the equality $\Delta = T \circ \Delta$ need only be verified on L, which generates $U(L)$ as an algebra over k. But $\Delta(x) = x \otimes 1 + 1 \otimes x = T \circ \Delta(x)$ for all $x \in L$.

(c) The foregoing observations, together with 2.2.13 (2), yield immediately:
For every connected commutative and cocommutative (graded) Hopf-algebra H there is a natural homomorphism of graded algebras $\varphi: U(P(H)) \to H$ which is induced by the inclusion $P(H) \subset H$ (φ is actually a homomorphism of Hopf-algebras). The main result is the following

Theorem: In the situation above (char $k = 0$!) we have: $\varphi: U(P(H)) \to H$ is an isomorphism

Proof. This is a particular case of [MM, 4.18,p.234] combined with [MM, 5.18,p.244]. Note that the standard (i.e. trivially graded) version of this theorem can be found in [BL, p.15].
The important fact is the possibility to reconstruct H as a free object over $P(H)$.

In order to tie finally everything together, we have to show the following

Proposition 2.2.15 k a field of characteristic 0, A a unital associative k-algebra, $H = H_*(gl(A))$ the homology of $gl(A)$, with its structure as a connected commutative and cocommutative (graded) Hopf-algebra (2.2.12(4)).
Then $P(H) = \mathrm{Prim}\, H_*(gl(A))$ (in the sense of 2.2.9)

Proof. Recall first the (adjoint) action of $g = gl_\infty(k)$ on $E = E(g \otimes A)$ via $\theta: g \to \mathrm{Der}_k E$, where

$$\theta(x)((x_1 \otimes a_1) \wedge \ldots \wedge (x_n \otimes a_n)) = \sum_{i=1}^{n} (x_1 \otimes a_1) \wedge \ldots \wedge [x, x_i] \otimes a_i \wedge \ldots \wedge (x_n \otimes a_n)$$

Considering the action of g on $E \otimes E$ given by $x \to \theta(x) \otimes 1 + 1 \otimes \theta(x)$, it is easy to verify that

$$\Delta\theta(x) = (\theta(x) \otimes 1 + 1 \otimes \theta(x))\Delta \quad \text{for all } x \in g$$

(i.e. the comultiplication Δ is a g-homomorphism)

Thus gE is a coideal of E:

$$\Delta(gE) \subset E \otimes gE + gE \otimes E$$

and $E(g \otimes A)_g$ becomes a quotient coalgebra of $E = E(g \otimes A)$.

But, revising the arguments about the well-definedness (and associativity) of the multiplication on $H_*(gl(A))$ (Lemma 2.2.11!), we see that everything works already well on $E(g \otimes A)_g$. Thus (look at 2.2.12(3) and (4)) $E(g \otimes A)_g$ is a connected commutative and cocommutative (graded) Hopf-algebra over k.

Furthermore, the differential of $E(g \otimes A)_g$ is a (graded) derivation:

For $u = u_1 \wedge \ldots \wedge u_p$, $v = v_1 \wedge \ldots \wedge v_q$ (think of equivalence classes) we have: $d(u \circ v) = (du) \circ v + (-1)^p u \circ dv$

This is easily verified:

$$u \circ v = \begin{pmatrix} u_1 & 0 \\ 0 & 0 \end{pmatrix} \wedge \ldots \wedge \begin{pmatrix} u_p & 0 \\ 0 & 0 \end{pmatrix} \wedge \begin{pmatrix} 0 & 0 \\ 0 & v_1 \end{pmatrix} \wedge \ldots \wedge \begin{pmatrix} 0 & 0 \\ 0 & v_q \end{pmatrix}$$

and $\left[\begin{pmatrix} u_i & 0 \\ 0 & 0 \end{pmatrix}, \begin{pmatrix} 0 & 0 \\ 0 & v_j \end{pmatrix}\right] = 0 \qquad 1 \le i \le p,\ 1 \le j \le q$

The explicit formula for d yields immediately the assertion. Thus $(E(g \otimes A)_g, d)$ is a differential graded Hopf-algebra. In order to prove the assertion of our proposition, we have to show

(1) $P(E(g \otimes A)_g) = \mathbb{P}E(g \otimes A)_g$ (in the sense of 2.2.6(3))

(2) $P(H_*(gl(A))) = H_*(P(E(g \otimes A)_g))$

Look first at (1).

Recall our reduced isomorphism (2.2.5(3))

$$(k[\gamma_n] \otimes A^n) \underset{\gamma_n}{\otimes} (\mathrm{sgn}) \xrightarrow{\sim} E_n(g \otimes A)_g$$

$$\sigma \otimes (a_1, \ldots, a_n) \otimes 1 \quad \to \quad \hat{E}^a_\sigma \bmod gE_n(g \otimes A)$$

(where $\hat{E}^a_\sigma = E^{a_1}_{1\sigma(1)} \wedge \ldots \wedge E^{a_n}_{n\sigma(n)}$)

which identifies

$(k[U_n] \otimes A^n) \underset{\gamma_n}{\otimes} (\mathrm{sgn})$ with $\mathbb{P}_n E(g \otimes A)_g$ (2.2.7)

Let us calculate $\Delta(\hat{E}^a_\sigma) = \sum_{p+q=n} \Delta_{p,q}(\hat{E}^a_\sigma)$ where $\Delta_{p,q}(\hat{E}^a_\sigma)$ is given by the following formula:

$$\Delta_{p,q}(\hat{E}^a_\sigma) = \sum_{(\rho)} \varepsilon(\rho)\, {}'E^{a\rho}_{\rho,\sigma\rho} \otimes {}''E^{a\rho}_{\rho,\sigma\rho},$$

the sum running over all $\rho \in \gamma_n$ which are increasing on $[1,p]$ and on $[p+1,n]$; we have exlicitely

$$'\hat{E}^{a\rho}_{\rho,\sigma\rho} = E^{a_{\rho(1)}}_{\rho(1),\sigma\rho(1)} \wedge \ldots \wedge E^{a_{\rho(p)}}_{\rho(p),\sigma\rho(p)} \bmod gE_p(g \otimes A)$$

$$''E^{a\rho}_{\rho,\sigma\rho} = E^{a_{\rho(p+1)}}_{\rho(p+1),\sigma\rho(p+1)} \wedge \ldots \wedge E^{a_{\rho(n)}}_{\rho(n),\sigma\rho(n)} \bmod gE_q(g \otimes A)$$

Now, it is easily seen that the following is true:

$$E^{a_1}_{i_1 j_1} \wedge \ldots \wedge E^{a_r}_{i_r j_r} \equiv 0 \bmod gE_r(g \otimes A)$$

whenever $\{i_1,\ldots,i_r\} \neq \{j_1,\ldots,j_r\}$ (and both sets have cardinality equal to r).

Thus we can simplify:

$$\Delta_{p,q}(\hat{E}^{a}_{\sigma}) = \sum_{(\rho)}{}^{*} \varepsilon(\rho)\, 'E^{a\rho}_{\rho,\sigma\rho} \otimes {''E}^{a\rho}_{\rho,\sigma\rho} \bmod gE_p \otimes gE_q$$

where the sum runs now over all $\rho \in \gamma_n$ which are increasing on $[1,p]$ and on $[p+1,n]$, and such that $\sigma\rho[1,p] = \rho[1,p]$, $\sigma\rho[p+1,n] = \rho[p+1,n]$. Note that the set of those admissible $\rho \in \gamma_n$ is in bijection with the set of all partitions (I,J) of $\{1,\ldots,n\}$ such that $|I| = p$, $|J| = q$ and such that $\sigma I = I$, $\sigma J = J$.

Now we can conclude:

$\Delta(\hat{E}^{a}_{\sigma}) = \hat{E}^{a}_{\sigma} \otimes 1 + 1 \otimes \hat{E}^{a}_{\sigma}$ if and only if $\Delta_{p,q}(\hat{E}^{a}_{\sigma}) = 0$ for all (p,q) with $p+q = n$, $p > 0$ or $q > 0$.

But this is equivalent to saying that there is no nontrivial partition (I,J) of $\{1,\ldots,n\}$ which is invariant for the action of σ on $\{1,\ldots,n\}$, which means precisely that $\sigma \in U_n$, the conjugacy class of $t = (1,2,\ldots,n)$.

Now we have to show (2)

By (1) and the argument in the proof of 2.2.10 (we have actually a projection onto the complex of primitive elements in $E(g \otimes A)_g$) we get:

$$H_*(P(E(g \otimes A)_g) \subset PH_*(gl(A)).$$

But primitive homology classes have primitive representatives. Thus we get equality, as claimed.

This finishes the last section of our lectures.

Comments on chapter II.

The skeleton character of cyclic cohomology for the construction of noncommutative de Rham homology (considered as a direct limit which inverts the suspension operator S; cf. [Co]) has his counterpart in theorem 1.1.18, for cyclic homology. This result of M. Karoubi's is one step of his program of find the appropriate range for his character maps from Quillen's K-groups to de Rham cohomology (lying in reduced cyclic homology; cf. [Kb]). Our second section deals with cyclic homology of commutative algebras in characteristic zero. The ideas and results are due to J.L. Loday and D. Quillen [L.Qu] and culminate in theorem 1.2.18. It was not difficult to replace all original spectral sequence arguments by simpler mixed complex patterns.
As to the second part, which attacks the "additive K-theory" aspect of cyclic homology, I have first collected all the matrial of invariant theory which enters implicitely or explicitely in the proofs of the last section. There I have tried to strip down the correspondence cyclic homology - Lie algebra homology to its bare essentials. Theorem 2.2.10 is due to J.L. Loday and D. Quillen. The very last section on the Hopf-algebra aspect of the result is sometimes sketchy, since a little bit non-thematic.

References to chapter II.

[An,L] André, M.: Méthode simpliciale en algèbre homologique et algè bre commutative. LNM 32. Heidelberg: Springer 1967

[An,B] André, M.: Homologie des algèbres commutatives. Berlin, Heidelberg, New York: Springer 1974

[Be] Behrens, E.A.: Ring Theory. New York, London: Academic Press 1972

[BL] Bourbaki, N.: Éléments de Mathématique. Groupes et Algèbres de Lie II, III. Hermann, Paris 1972.

[Bou] Bourbaki, N.: Éléments de Mathématique. Algebre. Ch.10. Paris... : Masson 1980

[C.E.],[Co] as in chapter I

[Fo] Fogarty, J.: Invariant Theory. New York: Benjamin 1965

[Gr] Green, J.A.: Polynomial Representations of GL_n. LNM 830. Berlin, Heidelberg, New York: Springer 1980

[H.St.] as in chapter I

[Hu] Humphreys. J.: Linear Algebraic Groups. Berlin, New York: Springer 1975

[Kb] Karoubi, M.: Homologie cyclique des groupes et des algèbres. Homologie cyclique et K-théorie algébrique, I et II. Homologie cyclique et régulateurs en K-théorie algèbrique. C. R. Acad. Sci. Paris, Série I, vol. 297 (1983) 381-384 et 447-450 et 513-516 et 557-560

[L.Qu.] as in chapter I

[Ma] Matsumura, H.: Commutative algebra, 2nd ed. London...: Benjamin/Cummings: 1980

[ML] as in chapter I

[M.M.] Milnor, J.W., Moore, J.C.: On the Structure of Hopf Algebras. Ann. Math. 81 (1965), 211-264.

Further References.

The reader will find a lot of references in [Co],but also in [Ka]. He should also consult

[Ca] Cartier, P.: Homologie cyclique: rapport sur les travaux récents de Connes, Karoubi, Loday, Quillen... Sém. Bourbaki, exp. 621, Février 1984

The natural first extension of the material exposed in these lectures would be

[Ka], already cited, and, I think,

[Gw] Goodwillie, Th.G.: Cyclic homology, derivations and the free loopspace. Topology, 24(2) (1985), 187-215

For those who are interested in noncommutative differential calculus (see II.1.1) and in the $\mathbb{Z}_2$-graded version of cyclic cohomology, I recommed

[Kt] Kastler, D.: Cyclic Cohomology within the differential Envelope. Preprint. CPT Marseille Luminy.

LIST OF SYMBOLS AND NOTATIONS

$E^2_{p,q}$ 3

$E^2_{p,q} \underset{p}{\Rightarrow} H_n$ 7

Tot (M) 13

${}^{I}F^{P}\mathrm{Tot}\,(M)$, ${}^{II}F^{P}\mathrm{Tot}\,(M)$ 13

${}^{I}E^2_{p,q}$, ${}^{II}E^2_{p,q}$ 14

$\mathcal{C}(A)$ 19

$HC_n\,(A)$ 20

$C^{\lambda}_{*}\,(A)$ 22

$H^{\lambda}_{*}\,(A)$ 22

$({}_{B}M, d)$ 24

$HC_{*}\,(M)$ 24

C (A) 33

$\mathcal{B}(A)$ 39

$\bar{C}\,(A)$ 43

$\bar{C}\,(A)_{red}$ 44

$\bar{H}_n\,(A)$ 45

$\bar{H}C_n\,(A)$ 45

$\mathcal{C}^{t}\,(A)$ 58

$HC^{n}\,(A)$ 60

$C^{*}_{\lambda}\,(A)$ 62

$H^{*}_{\lambda}\,(A)$ 62

$({}^{B}M, d)$ 64

$C^{t}\,(A)$ 65

$\Omega_1\,(A)$ 86

$\Omega\,(A)$ 90

$\Lambda\Omega\,(A)$ 94

$\Omega^1_{A/k}$ 94

$H^n_{DR}\,(A)$ 117

$[\underline{r}, \underline{n}]$ 121

S (n) 122

$H_{*}\,(g)$ 130

$(E\,(g), d)$ 130

gl (A) 133

Prim H_{*} (gl (A)) 144

P (H) 150

INDEX

approximation theorem 10
convergence (of a spectral sequence) 7
cyclic
    cohomology 60
    homology 20
de Rham
    cohomology 95, 117
    complex 94
    homology 96
differential envelope 90
d. g. algebra 25
double complex 12
enveloping algebra 17, 127
exact couple 1
    derived 3
filtration 1
    bounded 7
    first and second 13
Hochschild
    cohomology 59
    complex 18
    homology 18
Homology of Lie algebras 130
Hopf-algebra
    graded 145
    commutative and cocommutative 149
Kähler-de Rham complex 95
limit term (of a spectral sequence) 7
mixed
    complex 23
    cochain complex 63
Morita-equivalent 74
normalized Hochschild complex 42
primitive elements 141, 144, 150
reduced
    cyclic homology 45
    Hochschild homology 45
Schur algebra 122
shuffle product 105, 107, 110
spectral sequence 6
strongly homotopy Λ-map 30
total complex 12, 13